Seed Policy, Legislation and Law: Widening a Narrow Focus

Seed Policy, Legislation and Law: Widening a Narrow Focus has been co-published simultaneously as *Journal of New Seeds*, Volume 4, Numbers 1/2 2002.

The *Journal of New Seeds* Monographic "Separates"

Below is a list of " separates," which in serials librarianship means a special issue simultaneously published as a special journal issue or double-issue *and* as a "separate" hardbound monograph. (This is a format which we also call a "DocuSerial.")

"Separates" are published because specialized libraries or professionals may wish to purchase a specific thematic issue by itself in a format which can be separately cataloged and shelved, as opposed to purchasing the journal on an on-going basis. Faculty members may also more easily consider a "separate" for classroom adoption.

"Separates" are carefully classified separately with the major book jobbers so that the journal tie-in can be noted on new book order slips to avoid duplicate purchasing.

You may wish to visit Haworth's website at . . .

http://www.HaworthPress.com

. . . to search our online catalog for complete tables of contents of these separates and related publications.

You may also call 1-800-HAWORTH (outside US/Canada: 607-722-5857), or Fax 1-800-895-0582 (outside US/Canada: 607-771-0012), or e-mail at:

getinfo@haworthpressinc.com

Seed Policy, Legislation and Law: Widening a Narrow Focus

Niels P. Louwaars
Editor

Seed Policy, Legislation and Law: Widening a Narrow Focus has been co-published simultaneously as *Journal of New Seeds*, Volume 4, Numbers 1/2 2002.

CRC Press
Taylor & Francis Group
Boca Raton London New York

CRC Press is an imprint of the
Taylor & Francis Group, an informa business

Reprinted 2010 by CRC Press

CRC Press
6000 Broken Sound Parkway, NW
Suite 300, Boca Raton, FL 33487
270 Madison Avenue
New York, NY 10016
2 Park Square, Milton Park
Abingdon, Oxon OX14 4RN, UK

Published by

Food Products Press®, 10 Alice Street, Binghamton, NY 13904-1580 USA

Food Products Press® is an imprint of The Haworth Press, Inc., 10 Alice Street, Binghamton, NY 13904-1580 USA.

Seed Policy, Legislation and Law: Widening a Narrow Focus has been co-published simultaneously as *Journal of New Seeds*, Volume 4, Numbers 1/2 2002.

The development, preparation, and publication of this work has been undertaken with great care. However, the publisher, employees, editors, and agents of The Haworth Press and all imprints of The Haworth Press, Inc., including The Haworth Medical Press® and Pharmaceutical Products Press®, are not responsible for any errors contained herein or for consequences that may ensue from use of materials or information contained in this work. Opinions expressed by the author(s) are not necessarily those of The Haworth Press, Inc. With regard to case studies, identities and circumstances of individuals discussed herein have been changed to protect confidentiality. Any resemblance to actual persons, living or dead, is entirely coincidental.

Cover design by Jennifer M. Gaska

Library of Congress Cataloging-in-Publication Data

Seed policy, legislation and law : widening a narrow focus / editor, Niels P. Louwaars.
    p. cm.
    Includes bibliographical references and index.
    ISBN 1-56022-092-9 (hbk. : alk. paper)–ISBN 1-56022-093-7 (pbk. : alk. paper)
    1. Seed industry and trade–Law and legislation. 2. Plant varieties–Protection. 3. Seed industry and trade–Government policy. I. Louwaars, N. P.
K3881.2.S44 2002
338.1'8–dc21
                                                          2002006051

Indexing, Abstracting & Website/Internet Coverage

This section provides you with a list of major indexing & abstracting services. That is to say, each service began covering this periodical during the year noted in the right column. Most Websites which are listed below have indicated that they will either post, disseminate, compile, archive, cite or alert their own Website users with research-based content from this work. (This list is as current as the copyright date of this publication.)

Abstracting, Website/Indexing Coverage · · · · · · · · · Year When Coverage Began

- *AGRICOLA Database <www.natl.usda.gov/ag98>* **1999**

- *BIOBASE (Current Awareness in Biological Sciences)*
  <URL: http://www.elsevier.nl> **2000**

- *Biology Digest (in print & online).* **1999**

- *Bioscience Information Service of Biological Abstracts*
  (BIOSIS) <www.biosis.org> **2001**

- *BUBL Information Service. An Internet-Based Information*
  Service for the UK higher education community
  <URL: http://bubl.ac.uk/> **2000**

- *Cambridge Scientific Abstracts (Agricultural & Environmental*
  Biotechnology Abstracts) <www.csa.com> **2001**

- *CNPIEC Reference Guide: Chinese National Directory*
  of Foreign Periodicals. **1999**

- *FINDEX <www.publist.com>* **1999**

- *Food Science & Technology Abstract (FSTA) <www.ifis.org>* **1999**

(continued)

Special Bibliographic Notes related to special journal issues (separates) and indexing/abstracting:

- indexing/abstracting services in this list will also cover material in any "separate" that is co-published simultaneously with Haworth's special thematic journal issue or DocuSerial. Indexing/abstracting usually covers material at the article/chapter level.
- monographic co-editions are intended for either non-subscribers or libraries which intend to purchase a second copy for their circulating collections.
- monographic co-editions are reported to all jobbers/wholesalers/approval plans. The source journal is listed as the "series" to assist the prevention of duplicate purchasing in the same manner utilized for books-in-series.
- to facilitate user/access services all indexing/abstracting services are encouraged to utilize the co-indexing entry note indicated at the bottom of the first page of each article/chapter/contribution.
- this is intended to assist a library user of any reference tool (whether print, electronic, online, or CD-ROM) to locate the monographic version if the library has purchased this version but not a subscription to the source journal.
- individual articles/chapters in any Haworth publication are also available through the Haworth Document Delivery Service (HDDS).

Seed Policy, Legislation and Law: Widening a Narrow Focus

CONTENTS

SEED LEGISLATION AND COUNTRY CASES

UPCOMING ISSUES

ABOUT THE EDITOR

Niels P. Louwaars is Research Manager at Wageningen University and Research Centre in The Netherlands with special responsibilities regarding international cooperation. After his education as a plant breeder at Wageningen University, he worked in seed sector development programs in various developing countries. Upon his return to The Netherlands, he advised several governments and international organizations such as FAO, the World Bank, and international agricultural research institutes and seed associations on seed policies and legislation. He is a respected authority on aspects of seed system development and on the international agreements and laws that have an influence on national seed policy and legislation. For this volume, he has developed an international network of seed specialists, bringing together a variety of views on the key elements that form the basis of seed policies and legislation.

Preface

Seed is both an important vehicle for technology transfer that plays an important role in agricultural development, and at the same time it is a commercial commodity. Seed policies, therefore, have to combine multiple and often contradictory objectives.

Seed industry development has long been regarded as a linear process in time. Seed policies in most countries currently concentrate on stimulating private enterprise. More recently developed concepts of integrated seed systems indicate that parallel and integrated development of formal and farmers' seed systems is needed to increase seed quality and availability with the majority of farmers in developing countries. Moreover, technical innovation, such as in biotechnology, and international agreements such as those on intellectual property rights and genetic resources require policy makers dealing with seed systems to include a wide array of issues in their analyses and actions.

Policies, therefore, need to develop a much wider scope than before. The increased complexity also results in the notion that blueprint solutions to seed policies and regulation, that were common during and after the Green Revolution, cannot be effective in all situations. This volume is, therefore, not intended to develop one solution for all seed related policy questions. It is a compilation of issues, looked at from a wide range of perspectives.

This volume brings together authors with widely varying backgrounds (economists, anthropologist, agronomists, biotechnologists and seed system specialists) from different continents. Their active participation in the preparation of this book is gratefully acknowledged.

Also important for the inception of this volumes are those who stimulated me to dig deeper into the seed supply issues in developing countries, in particular Mr. S. H. Charles of the Ministry of Agriculture in Sri Lanka in the 1980s, Emmanuel Gareeba Gaso of the Uganda Seed Projects, Louise Sperling (CIAT), Rob Tripp (ODI), Conny Almekinders (Wageningen University) and Jaap Hardon. Their support in the development of new ways to look at formal

[Haworth co-indexing entry note]: "Preface." Louwaars, Niels P. Co-published simultaneously in *Journal of New Seeds* (Food Products Press, an imprint of The Haworth Press, Inc.) Vol. 4, No. 1/2, 2002, pp. xxi-xxii; and: *Seed Policy, Legislation and Law: Widening a Narrow Focus* (ed: Niels P. Louwaars) Food Products Press, an imprint of The Haworth Press, Inc., 2002, pp. xiii-xiv. Single or multiple copies of this article are available for a fee from The Haworth Document Delivery Service [1-800-HAWORTH 9:00 a.m. - 5:00 p.m. (EST). E-mail address: getinfo@haworthpressinc.com].

and local seed systems, on privatisation and management of agro-biodiversity, have been extremely important in the production of this volume. The contacts with the many people who have gained an interest in seed policies will continue to be an indispensible motivation to continue to support seed system development in the developing countries.

It is hoped that this book will contribute to the creation of effective seed policies and to the improvement of seed regulatory frameworks that will help farmers to better maintain or access the best seeds for their use.

Niels P. Louwaars

Seed Policy, Legislation and Law: Widening a Narrow Focus

Niels P. Louwaars

SUMMARY. Seed is a major technology transfer vehicle and the efficiency of seed supply systems to cater for the needs of different types of farmers is an issue in agricultural development policies. At the same time, seed is a (potential) commercial product and it is the carrier of valuable genetic resources. Seed policies have concentrated on the commercialisation aspects. More recently, however, the development value of seed for resource-poor farmers and the policies for *in-situ* conservation of plant genetic resources have created a more diverse interest in seed supply, especially farmers' seed systems and participatory approaches in breeding and seed provision. This creates the need to involve in the seed policy debate, not only conventional seed specialists, but also environmental specialists, (small) business development specialists and social scientists. The governments may take three different positions in the regulation of seed supply chains: competition, cooperation, control. In addition to domestic objectives, seed regulation is influenced more and more by international agreements, such as those on intellectual property, biosafety and business regulations among seed association members. *[Article copies available for a fee from The Haworth Document Delivery Service: 1-800-HAWORTH. E-mail address: <getinfo@haworthpressinc.com> Website: <http://www.HaworthPress.com> © 2002 by The Haworth Press, Inc. All rights reserved.]*

KEYWORDS. Seed policy, agro-biodiversity, farmers' seed systems, seed chain, regulatory reform

Niels P. Louwaars is affiliated with the Plant Research International, P.O. Box 16, 6700 AA Wageningen, The Netherlands.

[Haworth co-indexing entry note]: "Seed Policy, Legislation and Law: Widening a Narrow Focus." Louwaars, Niels P. Co-published simultaneously in *Journal of New Seeds* (Food Products Press, an imprint of The Haworth Press, Inc.) Vol. 4, No. 1/2, 2002, pp. 1-14; and: *Seed Policy, Legislation and Law: Widening a Narrow Focus* (ed: Niels P. Louwaars) Food Products Press, an imprint of The Haworth Press, Inc., 2002, pp. 1-14. Single or multiple copies of this article are available for a fee from The Haworth Document Delivery Service [1-800-HAWORTH 9:00 a.m. - 5:00 p.m. (EST). E-mail address: getinfo@haworthpressinc.com].

1

WHY SEED POLICY?

Seed is a primary input in agriculture next to land and labour. Seed has therefore been of primary concern to all farmers since the dawn of agriculture some 10,000 years ago. Only since the second half of the 19th century, however, seed has become an issue in agricultural policy development following the notion that seed quality and especially the genetics embedded in the seed are important determining factors for yield and product quality. This notion and the gradually increasing understanding of seed physiology from the first scientific publication on seed physiology by de Candolle (1832) resulted in the establishment of the first official seed testing laboratories in the 1880s. In most developing countries, seed became the subject of agricultural policies only during the Green Revolution, i.e., from the 1960s onward. Seed was seen as an important vehicle for the dissemination of technology, both the technology embedded in the seed itself (the short-straw characteristic of rice and wheat), and the technology that accompanied the new genes such as chemical fertilisers and plant protection chemicals.

In the industrialised countries, seed had turned by that time from a technology carrier to a commercial commodity.

Having such powerful characteristics for increasing yields that could contribute to feeding urban populations and bring prosperity in rural areas, the seed secured an important place on the national and international policy agendas. Worth mentioning is the Seed Industry Development Programme of FAO/UNDP that was to develop the infrastructure and technical know-how for mass dissemination of the new Green Revolution varieties to as many farmers as possible in a large number of developing countries. The development orientation led to the organisation of large government seed farms, contract grower schemes and government run seed conditioning facilities, shaped after successful seed companies in the North.

The further development of such initial seed multiplication facilities was thought to be linear throughout the world. Policies were simply needed to lift a national seed system to a higher level. Douglas (1980) was one of the most prominent originators of this historic approach. He developed the concept of 'natural' stages in seed industry development that should further the SIDP initiatives in the public sector towards a sustainable commercial seed industry. This concept was the basis of the analysis of seed policies by the International Food Policy Research Institute in 1991 (Pray & Ramaswami, 1991) where the following four stages were identified:

1. no seed industry because no improved varieties,
2. farmers start to use varieties from formal breeding but most seed is still produced by farmers,

3. introduction of private sector along with public enterprises, and
4. most seed purchased; bred by private research.

Policy options based on that analysis are quite straightforward. In order to move a seed industry from Stage 1 to 2, investments in breeding have to be made. In Stage 2, the quickest way to increase the use of 'improved' seed is to develop public seed multiplication centres. Stage 3 calls for measures to stimulate private investment in the seed industry and Stage 4 requires the banning of farmers' seed through education or regulation.

SEED POLICIES: COLLIDING INTERESTS

Increasing Complexity

Reality proved more complex than this very rational approach and alternative strategies for seed system development (Camargo et al., 1989) came up along with new concepts in technology development in agriculture (e.g., by Chambers et al., 1989). The compilation of a range of case studies on the complexities and values of farmers' knowledge on seed selection, saving and dissemination in different parts of the world resulted in an increased interest in traditional seed handling practices (Cromwell et al., 1993). In turn this led to the concept of integrated seed systems, whereby the strengths of both formal and local knowledge are combined in various operations within the seed system (Louwaars & van Marrweijk, 1994).

Currently this concept is furthered in terms of participation in breeding (Witcombe et al., 1996; Sperling et al., 1993) and participatory seed system development (Almekinders & Louwaars, 1999), small seed enterprise development (Kugbei et al., 2000) and various other forms.

The diversity of approaches makes it much more difficult to develop a comprehensive blueprint seed policy. This is the main reason for the compilation of this book, and to structure it in a number of papers rather than one comprehensive analysis of issues. Our main theme is that seed systems have to be understood in order to develop policies and regulations rather than establishing the development stage and acting accordingly.

Combining Different Functions

The two main functions of seeds create the first problem in seed policy development: the technology transfer function of seed and the ability of seed itself to be a commercial commodity.

The first function calls for a free availability of new varieties for distribution and dissemination from farmer to farmer. This means in most cases policies towards subsidised seed multiplication programmes based on public breeding and a limited scope of seed regulation (at best to stimulate quality-awareness with the public seed producers). This approach almost by definition kills private initiative for establishing a more efficient organisation of seed supply and adding the commodity function of seed.

Emphasis on the commercial aspects of seed may contradict development objectives. An example is the formal policy to make the public seed production system in Uganda commercially viable in 1990. After a thorough analysis of the cost-benefit ratio of the various seeds, the government seed enterprise suggested to concentrate on profitable maize, sorghum and groundnut seeds and to abandon bean seed production. Bean seed production in a centralised unit proved not a profitable operation due to the low multiplication factor and the bulkiness of the seed, increasing production costs. On the other hand, the ease with which farmers themselves could produce an acceptable seed quality on-farm quite cheaply, resulted in low seed prices in the market. The importance of beans in the local diet, however, and especially the value of the crop for the protein supply for the poorer sectors in the community resulted in the softening of the commercialisation policy.

A third function of seed has increased in significance over the past decades: seed as the carrier of valuable genetic resources that need to be conserved for future generations. Ideas for the conservation of genetic resources for a long time were monopolised by the genebank concept: collect, store (*ex situ*) evaluate, document and make available. The notion that this concept basically represents the conservation of the available diversity at a certain moment, and that after that resources can only be lost and not gained, led to concepts for 'in situ' and on-farm management of genetic resources. The latter concept is closely linked with maintaining and developing farmers' seed systems rather than replacing farmers' seed with uniform modern varieties. This further complicates seed policy matters since objectives to manage genetic diversity and objectives to increase agricultural production through breeding and the use of quality seed have to be merged.

Combining Different Policy Levels

Seed policies within the concept of linear seed industry development laid down by Douglas (1980) mentioned above were developed by a relatively small group of people, the seed specialists. These seed specialists were able to determine what was needed to move to the next stage of development. These determined in many countries that an independent seed quality control system was to be developed when the (public) seed units obtained a certain size and complexity. A next step was to call for a seed legislation that laid the legal foundation for such seed quality control institutions and that made their position clear vis-à-vis the seed producers and seed merchants. Even though there are opposing views and interests in these developments, the policy domain is narrow (Knoke et al., 1996).

From the early 1990s onwards, however, the domains have widened considerably due to two reasons:

1. Local seed systems were more widely acknowledged as important for the dissemination and sustainable use of new technology, and for the management of genetic resources. Already in the latter periods of the SIDP, so called lateral spread was considered important to reach remote and resource-poor farmers with new varieties. Only in the past decade, the local seed systems started to be considered valuable for their own creative capabilities (participatory plant breeding), and for alternative ways to increase the quality of the seed, used by farmers. This development brings in farmers and NGO's into the seed policy arena. Since the diversity within a country can vary considerably among countries, it is not possible to develop a blueprint model for seed policy development.
2. International developments influence the national seed policy agenda more and more. The economic agreements and in particular those under the World Trade Organisation (WTO) lead to pressures to liberalise markets, including the seed market. The TRIPs Agreement (agreement on Trade Related Aspects of Intellectual Property Rights) of the WTO requires countries to establish property rights systems for plant varieties, leads to debates about the required scope of protection. The International Treaty on Genetic Resources for Agriculture (IT/GRFA) of the FAO introduced the concept of farmers' rights, and the Convention on Biological Diversity (CBD) the national sovereignty over genetic resources. Also these concepts influence seed policies at the national level. In most countries this removes the mandate of seed policies from the ministry responsible for agriculture to the cabinet, because where the FAO falls under the manadate of that ministry, the CBD deals with Ministries of Environment and the WTO with those for Commerce (Economic Affairs). These ministerial responsibilities are regulated by often very different policy actors and networks.

Especially where it comes to translating policies into practice, e.g., through legislation, the widening of the policy arena makes the process much more complex, and subject to changing power relations. Blueprint models cannot be developed and trends are to develop policies for sub-issues, for example, for seed enterprise development (Tripp & Rohrbach, 2001). A good example of the struggle to bring such sub-policies together in one legal framework is the plant variety protection act in Thailand. Different classes of varieties are identified in order to comply with diverse policy objectives, e.g.,

- Stimulate (private) breeding through intellectual property rights,
- Guaranteeing wide availability of the public varieties,
- Stimulate on-farm conservation of agro-biodiversity through farmers' rights.

This illustrates very well the challenge that interdependency of functions and actions poses to decision-makers. Modeling of such processes has not been attempted yet, but the growing complexity of issues may warrant the introduction of such methodologies as described by Rajabi et al. (1997). Nijkamp (1998), however, argues that optimisation models have severe limitations where a common denominator for the different objectives (e.g., money) cannot be determined. The OECD has developed its Regulatory Impact Analysis in order to tackle the problem of integration of multiple policies, especially those concerning economic efficiency, trade, equity and environment. This model not only creates an analytical model, but is also meant to improve transparency of the policy-making processes, and accountability of the policy makers. Lack of transparency domain remains a major bottleneck in most countries, resulting in a sub-optimal use of the knowledge and views of different actors in the policy domain.

OPTIONS FOR POLICY IMPLEMENTATION

The "3 Co's": Competition, Cooperation, Control

Translating broad policy objectives into action depends to a very large extent on differing philosophies about the role of the government in society. The three broad options are summarised in "the 3 Co's": Competition, Cooperation, Control. These options can be illustrated very well in terms of regulating the seed sector.

Control

The fact that the farmer (buyer) can hardly ever determine the quality by looking at the seed, governments have developed systems of consumer protection through seed certification and quality control. The level of government involvement is a very important distinguishing factor between countries.

Most European countries (both east, central and west) adopted a policy best described with "Control." Seed regulatory framework gives government a strong voice in different stages of the seed chain: new varieties have to be registered, and undergo government-run multilocational variety tests for their Value for Cultivation and Use (VCU) before being formally released. The seed production chain is controlled through a certification system. Official seed testing and some marketing control finally closes the consumer-protection system (Figure 1).

FIGURE 1. The formal seed chain (ellipses) and its public control mechanisms (rectangles)

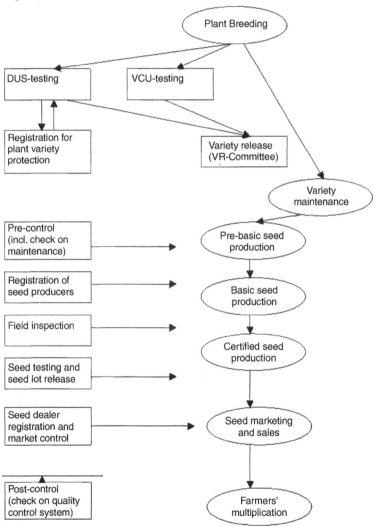

Apart from consumer protection this system is meant to create a level playing ground for seed enterprises to compete without undue competition from 'fly-by-night' seed suppliers.

Control-options very often appear in the process of privatisation of the seed industry in developing countries. When foreign companies show an interest in entering a country's seed system, governments get wary of protecting their farming community through seed controls. This in turn may scare-off the investors, thus missing a chance to create a more competitive market.

Competition

The other extreme is to fully rely on market forces. In this philosophy, competition in the seed market is the only sustainable factor for securing optimal seed quality in the market, and thus for consumer protection. This approach is not implemented in any western country, even though the argument is widely used in discussions about seed regulatory reform, especially by American (US) seed specialists. The concept is very straightforward: in a competitive seed industry, suppliers will arrange for all seed quality control operations within their companies; supplying poor quality seed will automatically result in a sever loss of market share.

This approach of no or voluntary regulations, assumes full competition in fully transparent markets. In practice this is hardly ever the case in seed markets in developing countries. Seed markets in such countries are often too small for a large-enough number of seed companies to operate, and the access to information is skew, with suppliers and large-scale farmers having the advantage.

Even though economic policies in the USA (and those promulgated in the WTO) do have confidence in market forces, the first laws that were to control seed quality was passed in the USA in 1905 (Copeland & MacDonald, 1979), followed by a comprehensive federal seed act in 1936.

Cooperation

The third option for a government is to design ways to cooperate with seed industry and to share tasks and responsibilities. The US-approach fits within this general approach: the seed supplier is fully responsible for the quality of its products, both in terms of adaptation of the variety to the client's agro-climatic conditions and in terms of seed quality. The government, however, has a strong grip on the type of information that the seed dealer should supply on the label, and on checking the truth-in-labeling. It considers that the customers (farmers) are sufficiently educated to determine which seed quality they want to buy, and to determine which suppliers are sufficiently honest in their promotion messages. Furthermore, different states have strong legislation on particular aspects on seed quality, with particular emphasis on contamination with noxious weed seeds.

A wide range of seed certification and seed quality control agencies exist in the USA that originate from farmers' and seed growers' associations. They perform the same functions as their counterparts in Europe, but the legal basis of their operations is weaker compared with their colleagues across the Atlantic.

Also in a more control-oriented seed legislation, cooperation can be developed that increase the effectiveness and efficiency of the control systems. Seed certification and control agencies in Europe may be run by the government (Germany) or by an independent foundation (The Netherlands), that originates from farmers' or seed associations. In the latter option, some levels of bureaucracy can be avoided. Within their operation, certification agencies may perform a full control as depicted in Figure 1, or they may concentrate on certifying the internal quality control operations of the seed companies, rather than certifying each and every seed lot throughout the chain. This option can be found in Europe especially in the horticultural seed sector, where the market is much more volatile than the more conservative field crops sector. Also in the field of variety testing for VCU and for DUS (Plant Variety Protection), there is a trend to use the facilities and the data of the applicant more in the examination of the value of the varieties.

TRENDS IN SEED LEGISLATION IN DEVELOPING COUNTRIES: DEREGULATION

Regulatory Reform

The first seed laws that were developed in most developing countries were based on the blueprint model supplied by the Seed Industry Development Programme. The details depended to a large extent on the origin of the advisors that assisted in the development of the seed sector. India adopted a US-type truth-in-labeling system, but combined this with a very strong European-type variety controls system. The seed laws of many other countries simply took either the examples of Spain, England or France as a basis, depending on their colonial background and official language (Bombin-Bombin, 1980). While taking these European laws as a basis, they often picked the most controlling examples from these countries, i.e., variety release based on government-run multilocational testing, full certification and seed testing of all crop seeds (Tripp & van der Burg, 1997; Tripp & Louwaars, 1997b). The level of regulation was commonly not in line with the level of institutional development, leading to incomplete implementation and insufficient transparency. This creates a serious lack of credibility to the system and possibilities for officials to apply certain regulations at will.

The reform of these standard regulatory systems to the needs of changing seed supply has recently been researched by the Overseas Development Institute (Tripp, 1997; Tripp & Louwaars, 1997b). The reform of these systems depends to a large extent on the existing institutions and organisations and bureaucratic considerations, on regulatory cultures and international pressure. The latter is more diverse than in previous decades. Formal pressures relate to deregulation in the framework of globalisation and liberalisation of markets, and the introduction of genetic resources aspects in seed policies. Moreover, similar to the influence that development assistance projects had for the development of the public seed enterprises and their regulatory basis, recent projects put the needs of farmers' seed production and small enterprise development at the political agenda. This introduces national and international NGOs into the policy debate.

Both pressures lead to calls for deregulation. Liberalisation of trade requires abandoning of hindrances by inefficient control agencies, while disregarding the risks associated with lifting controls in low-value markets, and with the use of poor quality seed by small and medium level farmers. Also small seed enterprise development may be helped with the lifting of government controls, especially where the cost of the inefficiencies of such institutions are translated into the certification and testing fees.

Both commercial seed companies that want to introduce varieties from their international breeding programmes and local seed enterprises and farmer-groups that may want to provide seed on locally adapted (landrace) varieties would benefit from reduced variety controls. Large seed companies may like to work with a (ISO) certification of their own quality control system. Also, small scale seed providers would better thrive on brand name than on government approval, especially where the regulatory institutions do not have the capacity to efficiently check on the often remote seed production fields.

Proposals to deregulate and to turn compulsory regulations into voluntary ones (Gisselquist, pers. comm.), however, have to go hand in hand with education of seed users where many farmers are unable to read seed labels. A truth-in-labelling system can very well be combined with certain minimum standards in order to reduce the risks of farmers.

Developments That Oppose Deregulation Trends

Deregulation is a globally accepted trend in many sectors. The SPS/TBT-agreement of WTO prescribes a framework for acceptable regulations for non-tariff trade barriers, which do not include extensive seed controls other than strictly necessary phytosanitary measures. There are, however, tendencies that go the opposite direction. These are particularly intellectual property rights, biosafety and access regulations on genetic resources.

The WTO-TRIPs Agreement spurred the development of plant variety protection regulations in many developing countries. Opposite to the trends in conventional seed legislation that intend to abandon variety registration, the implementation of such variety protection laws does require a detailed registration of the protected subject matter. Moreover, national and international seed trade will be influenced by protection claims, not so much because of government laws, but as the result of private law contracts and private market control organisations, such as ARPOV in Argentina, will regulate the seed trade. Public agencies that intend to make their varieties available for the public without claiming rights are likely to register their varieties as well, if only to avoid others from claiming ownership and restrict the use to their licensees only.

A similar trend that private contracts influence the seed trade can be observed in countries where patent-protected transgenic crops are planted at a considerable scale.

Similarly, the introduction of transgenics triggers more and more countries to develop biosafety regulations that make the release of these varieties both for testing and for commercial use, dependent on release procedures. In case the use of such varieties becomes more widespread, the reduction in variety registration procedures due to deregulation in conventional seed laws will be replaced by comprehensive release procedures under biosafety laws.

Thirdly, new developments within the framework of CBD and IU/GRFA may lead to even more variety regulatory systems than ever experienced under conventional seed laws. The Prior Informed Consent (PIC) system under CBD may lead to extensive variety regulations, and the farmers' rights principle may also lead to additional administrative procedures.

THIS VOLUME

This volume assumes that a blueprint approach to national seed policies is not useful. It is a compilation of views rather than an attempt to develop a work-book for seed policy makers that present the answers to all questions.

The section following this chapter deals with policy aspects in the seed sector. Almekinders and Louwaars (this volume) lay a conceptual foundation for identifying the roles of formal and farmers' seed systems, showing that policies and regulations that focus on the formal sector alone, are likely counter productive for increasing seed quality and availability for the majority of farmers in most countries.

Reusché, Kugbei and Bishaw, and van der Meer deal the important aspect of private enterprise development from widely different angles. It provides an insight in the importance of the institutional and market conditions within which enterprises are to be developed. Policies to develop small seed enterprises may not be the same as those for creating an optimal business environment for privatised public enterprises. The limitations of the current deregulation trend are analysed in the perspective of market failure.

Le Buanec and Heffer point at the importance of international associations in the creation of policies that support enterprise development.

Finally, Louwaars and co-authors illustrate the importance of technological innovation in policy development. This section illustrates that seed policy development is a continuous process of responding to changing perceptions about seed system development, and to changing economic and technological environment. It shows that reality is more complex than the existing linear models of seed system development, and that policies to guide this process cannot follow a blueprint model.

The following section deals with seed regulations. A general overview is presented by Tripp, van Gastel and co-authors, and Louwaars. These papers deal with conventional seed regulatory issues, such as seed certification and variety controls, and aspects of regulatory change. Le Buanec then describes the additional regulations of the international seed industry that receive little attention in seed policy analysis.

Three country papers illustrate these aspects. Three countries have been selected on the basis of their widely different farming systems, seed markets, and seed industry development histories: Turkey, Uganda, and Bangladesh.

This section shows that regulatory options that may seem suitable to support certain processes may contradict other useful developments and that policy makers need to have a wide enough focus to avoid unnecessary contradictions.

The final section deals with three policy fields that have an influence on seed policy and legislation in some countries, but that will have a significant impact on seed system development throughout the world. Traynor and Komen present recent developments in biosafety policies and their influence on variety registration practices. Visser highlights international developments in the field of access and benefit sharing of plant genetic resources that will certainly influence formal variety registration and possibly even local seed systems. Ghijsen finally describes different intellectual property rights systems and their influence on seed availability.

This section does not intend to go into all details of the three broad issues–this would warrant three new volumes–but it intends to highlight the importance of policy makers to widen their focus beyond the narrow sector of seed production and supply.

This volume is not intended to present solutions to all questions that policy makers in the seed sector face throughout the world. It does, however, combine a number of overview articles and analyses which highlight major topics that touch at the heart and the perimeter of the seed policy development process. It is hoped that this book will contribute to the development of effective seed policies and to the improvement of seed regulatory frameworks that will help farmers to better maintain or access the best possible seeds for their own agro-ecological and socio-economic conditions.

REFERENCES

Almekinders, C.J.M. & W.S. de Boef (eds.), 2000. *Encouraging diversity. Plant genetic resource conservation and crop development.* London, Intermediate Technology Publication.

Almekinders, C.J.M. & N.P. Louwaars, 1999. *Farmers' seed production, new approaches and practices.* London, Intermediate Technology Publications.

Bombin-Bombin, L. 1980. *Seed legislation.* Rome, Food and Agriculture Organisation.

Chambers, C.P., C. Bragantan & A.A Morales. 1989. *Seed production systems for small farmers: A non-conventional perspective.* CIAT, Cali, Colombia.

Chambers, R., A. Paceye & L.A. Thrupp. 1989. *Farmer first: Farmer innovation and agricultural research.* London, Intermediate Technology Publication.

Copeland, L.O. & M.B. McDonald, 1979. *Principles of seed science and technology* 2nd Ed. Minneapolis, Minnesota, Burgess Publication Co. pp. 279-289.

Cromwell, E., S. Wiggins & S. Wentzel, 1993. *Sowing beyond the State. NGOs and seed syupply in developing countries.* London, Overseas Development Institute, 143 p.

De Candolle, A.P. 1932. *Conservation des graines.* Physiologie Végétale Vol. 2 pp. 618-626. Paris. Quoted in Justice & Bass, 1979.

Douglas, J., 1980. *Successful seed programmes. A planning and management guide.* Boulder, CO, Westview Press.

Justice, O.L. & L.N. Bass, 1979. *Principles and practices of seed storage.* Castle House Publication Ltd., 289 p.

Knoke, D., F.U. Pappi, J. Broadbent & Y. Tsujinaka, 1996. *Comparing policy networks.* Cambridge Studies in Comparative Politics, Cambridge, New York, Melbourne, Press Syndicate, University of Cambridge, 288 p.

Kugbei, S., M. Turner & P. Witthaut (eds.), 2000. *Finance and management of small-scale seed enterprises.* Proceedings of a workshop Addis Ababa, Ethiopia, 26-30 October, 1998, Aleppo, Syria, ICARDA.191 p.

Louwaars, N.P. & G.A.M. van Marrweijk, 1994. *Seed supply systems in developing countries.* Wageningen, CTA.

Nijkamp, P., 1998. *Globalisation and localisation in sustainable development scenarios.* Nagoya, Japan, PRSCO Summer Institute. Abstract at <http://prsco.agbi.tsukuba.ac.jp/Meetings/abst_sum_5/Nijkamp.html>.

OECD Secretariat, 1996. *An overview of regulatory impact analysis in OECD countries.* Paris, OECD <http://www.ocde.org/puma/regref/reg-96-7.htm>.

Pray, C.E. & B. Ramaswami, 1991. *A framework for seed policy analysis in developing countries.* Washington, DC, International Food Policy Research Institute, 42 p.

Rajabi, S., D.M. Kilgour & K.W. Hipel, 1997. *Modeling action-interdependence in multicriteria decision making. European Journal of Operational Research* 110: 490-508.

Sperling, L., M. Loevinsohn & B. Ntambovura, 1993. Rethinking the farmer's role in plant breeding: local bean experts and on-station selection in Rwanda. *Experimental Agriculture* 29: 509-519.

Tripp, R. 1997. The dynamics and seed policy change: state responsibilities and regulatory reform. In: R. Tripp (ed.) *New seeds and old laws. Regulatory reform and the diversification of national seed systems.* London, Intermediate Technology Publication, pp. 157-173.

Tripp, R. & W.J. van der Burg, 1997. The conduct and reform of seed quality control. In: R. Tripp (ed.) *New seeds and old laws. Regulatory reform and the diversification of national seed systems.* London, Intermediate Technology Publication pp. 121-154.

Tripp, R. & N.P. Louwaars, 1997a. The conduct and reform of crop variety regulation. In: R. Tripp (ed.) *New seeds and old laws. Regulatory reform and the diversification of national seed systems.* London, Intermediate Technology Publication, pp. 88-120.

Tripp, R. & N.P. Louwaars, 1997b. Seed regulation: choices on the road to reform. *Food Policy* 22: 433-446.

Tripp, R. & D. Rohrbach, 2001. Policies for African seed enterprise development. Food Policy 26: 147-161.

Witcombe, J.R., A. Joshi, K.D. Joshi & B.R. Sthapit, 1996. Farmer participatory crop improvement. I. Varietal selection and breeding methods and their impact on biodiversity. *Experimental Agriculture* 32: 445-460.

SEED POLICY

The Importance
of the Farmers' Seed Systems
in a Functional National Seed Sector

Conny J. M. Almekinders
Niels P. Louwaars

SUMMARY. The farmers' systems of seed supply and crop development form by far the most important source of seed in most farming systems of the world. Despite the efforts of large seed programmes to replace the farmers' seed systems for a system in which farmers use seed as an external input, the major part of agricultural land in the world is still sown with seed that is informally produced by farmers. Aiming for a formal seed sector that supplies 100% of the seed for planting is only realis-

Conny J. M. Almekinders is affiliated with the Department of Technology & Agrarian Development, Wageningen University, Wageningen, The Netherlands.

Niels P. Louwaars is affiliated with the Plant Research International, Wageningen, The Netherlands.

Address correspondence to: Niels P. Louwaars, Plant Research International, P.O. Box 16, 6700 AA Wageningen, The Netherlands.

[Haworth co-indexing entry note]: "The Importance of the Farmers' Seed Systems in a Functional National Seed Sector." Almekinders, Conny J. M., and Niels P. Louwaars. Co-published simultaneously in *Journal of New Seeds* (Food Products Press, an imprint of The Haworth Press, Inc.) Vol. 4, No. 1/2, 2002, pp. 15-33; and: *Seed Policy, Legislation and Law: Widening a Narrow Focus* (ed: Niels P. Louwaars) Food Products Press, an imprint of The Haworth Press, Inc., 2002, pp. 15-33. Single or multiple copies of this article are available for a fee from The Haworth Document Delivery Service [1-800-HAWORTH 9:00 a.m. - 5:00 p.m. (EST). E-mail address: getinfo@haworthpressinc.com].

tic for a small number of crops and in few countries. The importance of farmers' seed systems merits that closer attention be paid to farmers' seed production and seed exchange at the policy level and in technical assistance projects. Linking formal and farmers' seed systems and improving the latter may in many cases be a more effective strategy to improve national and local seed supply than aiming only at improving the infrastructure and investment climate for the formal (private and public) seed sector. In fact, analysis of strengths and weaknesses of both the farmer and formal seed system shows important complementarity in strength and weaknesses between the two systems, which offers multiple opportunities for improving the effectiveness of both. Very few countries have included such an approach in their seed policies yet. This paper presents the importance of the farmers' seed systems from a variety of perspectives. We indicate ways for further integration of the formal and farmers' systems at various points in the seed chain/seed cycles and propose to include such strategies in national seed policies. *[Article copies available for a fee from The Haworth Document Delivery Service: 1-800-HAWORTH. E-mail address: <getinfo@haworthpressinc.com> Website: <http://www.HaworthPress.com> © 2002 by The Haworth Press, Inc. All rights reserved.]*

KEYWORDS. Farmers' seed systems, formal seed system, participatory plant breeding

INTRODUCTION

Farmer-produced seed is the most important source of planting materials in this world. Seed and other propagation materials that are produced on-farm and are not part of formal seed sector arrangements, e.g., contract production, is usually associated with farmers' or local seed. All of activities related to farmers' seed production and supply are commonly referred to as traditional (Cromwell et al., 1992), local (Almekinders et al., 1994; Louwaars & van Marrewijk, 1996), or farmers' seed systems (Almekinders & Louwaars, 1999). In this volume, we use the term farmers' seed system.

The importance of farmer-produced seed varies between crops, farms, regions and continents. It is, however, most important for small-scale farmers in low input agriculture in developing countries. Depending on the crop and country, 60-100% of the seed planted in developing countries is farmer produced and exchanged. When no formal sector breeding or seed supply exists, such as for indigenous vegetables and root crops like yam and sweet potato, farmers' seed is usually the only source of planting material. This is particular the case for local varieties and minor crops. Related with the importance of farmers' seed is the maintenance of varieties by farmers, including both im-proved and local varieties.

Farmers' seed systems also have a wider significance than the local supply of seed and maintenance of varieties, constituting a dynamic *in situ* conserva-tion system in which evolution continues to exist. This gives farmers' seed sys-tems an important role in the global management of Plant Genetic Resources for Food and Agriculture (PGRFA) (FAO, 1996).

In general, farmers' seed systems have not been considered by the formal seed sector and policy makers as a part of the seed sector. This article ad-dresses the character and importance of the farmers' seed systems and focuses principally on these systems in developing countries. The farmers' system is described as an integrated system that functions parallel to the formal system. It is pointed out that the farmers' and formal systems are poorly connected, while their complementarity indicates that a better integration offers opportu-nities to improve seed system functioning for all parties.

FARMERS' SEED SYSTEMS AND THE DEVELOPMENT OF A FORMAL SEED SECTOR

The existence of a farmers' seed system next to a formal system is the result of specialisation of seed production and breeding, which started in the early 20th century. After the re-discovery of Mendels' findings, genetics developed into a science that found its practical application in plant breeding. Specialisa-tion of plant breeding and the associated production of the seeds of improved varieties developed from farm-based enterprises into commercial companies. The research and basic breeding work became more complex with rapid ad-vances in genetics and increasingly carried out in public research institutions and universities. The development of a commercial breeding-seed sector was especially enhanced by the discovery of the phenomenon of heterosis and the generation of hybrid varieties (Kloppenburg, 1988). In industrialised coun-tries, this development was associated with an increased use of inputs. Higher levels of fertiliser use, chemical crop protection, mechanisation and a growing size of farms asked for input-responsive varieties that germinate, develop and mature uniformly. The development of commercial seed sector found its basis in the specialisation of farm production and economies of scale of seed pro-duction; it was a development that benefitted all parties. Seed laws were the re-sult of pressure from both farmer and seed producers, to provide protection against malafide seed suppliers, which affected farmers and the integrity of bonafide seed producers.

In developing countries, the developmental scenario of the seed sector has been different. In most countries, crop improvement did not receive a signifi-cant boost until the discovery and use of the dwarfing genes in the major food cereals. With the commercial development of maize hybrids in Zimbabwe and Kenya as exceptions, the first significant genetic improvements in wheat and rice occurred in most countries as a result of special projects initiated by the Rockefeller and Ford Foundations. A maize breeding programme in Mexico followed shortly thereafter. These first crop improvement initiatives laid the foundation of the Internationad Agricultural Research Centers and the Consultative Group of International Agricultural Research (CGIAR) system. Breeding, however, concentrated on the major food crops and on large 'recommendations domains,' aiming at reaching as many beneficiaries in developing countries as possible. This approach resulted in a bias towards high potential areas and was based on an approach replacing local materials with 'high yielding varieties.'

The products of these early breeding programmes were so successful in raising crop yields and ushering in the era of 'Green Revolution.' The adoption of these Green Revolution varieties has, however, not been as widespread as envisioned. The impact on the poor people in developing countries was questioned as well, because the differential effects on food prices and labour in many cases did not improve living conditions of the poorest in our society (Lipton and Longhurst, 1989; Kerr and Kolavalli, 1999). Green Revolution varieties were most successful in the areas that were characterised by relatively favourable and uniform agro-ecological and socio-economical conditions. In particular, the modern rice varieties generated by IRRI were widely adopted in the irrigated-rice systems in South East Asia. In other crops, the adoption of modern varieties had been more variable, with significant differences between crops, regions and continents (Table 1). It is important to realise that this relative success of the breeding efforts of the Green Revolution for the main staple

TABLE 1. Percent area planted to modern varieties of rice, wheat, and maize in developing countries (adapted from Byerlee, 1996)

	Rice (1983)	Wheat (1990)	Maize (1990)
Sub-Saharan Africa	15	52	43
West Asia/North Africa	11	42	53
Asia (excl. China)	48	88	45
China	95	70	90
Latin America	28	82	46
All developing countries	59	70	57

crops was not translated in a simlarly high percentage of seasonally supplied seed by the formal seed sector in these countries.

Despite the considerable effort of large scale seed programmes, often financed by the World Bank and UNDP, and supported by FAO in its Seed Industry Development Programme (FAO, 1994), the impact of these programmes has in general been below expectation (Kerr and Kolavalli, 1999). A blueprint for the development of a seed sector in developing countries following the stages of development of a seed industry in industrialised countries (Douglas, 1980) had a very important influence on seed policies in the South. Nevertheless, the seed programmes in the developing countries proved to be ineffective in a number of closely interacting points that relate to the characters and quality of distributed improved varieties and seeds, and the organisation of seed production and marketing. As a consequence, the public seed sector has struggled with its economic justification and organisation. Seed policies that developed in the 1980s followed the general economic policies of structural adjustment. Transformation of the public seed units into viable seed enterprises became the leading policy objective, which proved to be too difficult in most countries. For those crops in which such viable enterprise development was feasible, a commercial sector has indeed emerged. In most countries, a commercial sector exists for modern varieties of maize, and some important cash crops and vegetables. As a result, the area planted with seed supplied by the formal seed sector varies between crops and countries, which is still relatively small (Table 2). The present situation is that, there does not exist a clearly defined strategy for national seed sector development for other important food and minor crops in developing countries.

In the following, we describe the seed requirements of small-scale farmers in developing countries and reflect on these requirements against the properties of seed supply from the formal sector.

SEED SUPPLY AND SMALL SCALE FARMERS IN DEVELOPING COUNTRIES

An important characteristic of small-scale farming is the need for crop genetic diversity. The formal seed sector has difficulty in meeting such high levels of diversity. Diversity of crop genetic resources has two vital functions for farmer households: (i) it serves multiple purposes of consumption, use and marketing, and (ii) it enables farmers to cope with the variable, unpredictable environment and market conditions (Clawson, 1985; Bellon, 1996; Brush, 1986). These functions are particularly important for farmers in complex, diverse and risk-prone environments and who are largely subsistence farmers.

Consumption and other purposes. Crop products are used by the household in many different ways. Maize is utilised in South America, for example, as

TABLE 2. Areas sown with seeds supplied by the formal seed sector (public and private) in various crops and countries

	Formal	Farmers'	Year (Reference)
Rice			
Tanzania	1	99	1985 (DANAGRO, 1888)
Pakistan	6	94	1995/96 (Bishaw & Kugbei)
Egypt	38	62	1997/98 (NN 1999)
Turkey	28	72	(Kutay, 1997)
Beans			
Malawi	4	96	1985 (DANAGRO, 1888)
Zambia	12	88	1985 (DANAGRO, 1888)
Honduras	2	98	1990 (Corrales et al., 1991)
Egypt (faba bean)	14	86	1997/98 (Seed Sector Programme)
Maize (*)			
Zambia	19	81	1997 (Aquino et al., 1999)
Zimbabwe	70	30	1997 (Aquino et al., 1999)
Pakistan	36	64	1997 (Aquino et al., 1999)
Egypt	36	64	1997/98 (Seed sector Programme)
Honduras	30	70	1990 (Corrales et al., 1991)
All crops (average)			
Netherlands	75	25	(Ghijsen, 1996)
Germany	50	50	(Ghijsen, 1996)
Greece	10	90	(Ghijsen, 1996)

(*) assuming a replacement rate of 0.25 for improved O.P. varieties

staple food and for the preparation of special dishes and beverages while the husk leaves are used for the preparation and packaging of specialty food (e.g., 'humitas'), and stems are used as fodder and fuel. Ensete, widely grown by farmers in Southern Ethiopia, produces starch for consumption, while particular varieties have important medicinal values. The leaves of Ensete are used for bedding, roofing, fibre and fodder (Admasu and Struik, 2000). People in Irian Jaya grow special sweet potato varieties for particular traditional dishes and others for baby food. Young sweet potato leaves are eaten as a vegetable, whereas older harvested vines are used as pig feed (Prain et al., 2000). In the Philippines, red-grain rice is served for special meals and deserts are prepared with glutinous rice. In the Andes, native potato varieties are often used for gifts and as payment for the contract labour; improved varieties are grown for the market (pers. observ., CA). Farmers in Mexico sometimes plant red-seeded maize varieties in the border of maize fields to 'guard the crop' (Hernandez, 1985).

Environmental variation. Farmers in the Andes plant different maize varieties and potato mixtures in their fields at different altitudes to match the heterogeneous and variable conditions. In Sierra Leone, farmers grow a range of rice varieties to match the different soil conditions along the slope of their fields (Richards, 1987). Under highly variable rainfall and fluctuating pest and disease pressures, genetically heterogeneous local varieties and mixed cropping are often more stable yielding than genetically uniform varieties and mono-cropping. For example, farmers in Niger deal with the unpredictability of rainfall and soil moisture by using mixtures of different millet types (Brouwer et al., 1993). Finally, fluctuating prices for inputs and products are an important source of socioeconomic variation and a reason for farmers to diversify crop production.

The adoption of improved varieties from the formal sector in developing countries is variable. In Ethiopia, few improved varieties of any crop are planted, whereas in Zimbabwe, improved maize varieties are planted by small-scale farmers in areas where formerly local sorghum was growing. In Sierra Leone and India, improved rice varieties are grown on a large scale in many lowland rice systems, whereas in upland rice areas, almost exclusively local varieties are planted. In the Andes, small-scale farmers plant both improved and local potato varieties; the improved ones tend to be grown at lower altitudes for the market, and the local dominate at higher altitudes for home consumption and local use. The adoption of improved varieties has been much discussed. Although a generalisation of non-adoption of improved varieties to the preferences of small scale farmers' conditions holds true in many situations, it also has to be recognised that improved varieties have reached many farmers in those marginal areas and are forming an important addition to the farmers' local genetic diversity. Although developed for high to moderate input levels, improved varieties may also outyield local varieties in low-input conditions. Genetically heterogeneous local varieties may be stable yielders, but they are also in many cases low yielders and often lack resistance to new pests and diseases.

In comparison to the highly diversified farmers' needs, formal breeding and seed programmes distribute low levels of genetic diversity. Breeding programmes, although working with high levels of genetic diversity at the early cycles of selection, usually produce not more than 1 or two genetically uniform varieties for official release. The number of varieties that are maintained and distributed through seed programmes or commercial companies is also modest in comparison to the portfolio of materials used in the farmer systems. The commercial sector is mostly involved in open pollinated (O.P.) or hybrid varieties of cross-fertilising crops like maize, sorghum, vegetables and some other cash crops like potato, soybean and cotton. Where improved varieties of self-pollinating food crops were adopted, the farmers' system soon became an important source of seed as well. This is not only the case in developing countries, but also in North America and Europe, farmers' seed saving is important.

Even in Eastern Europe and the former Soviet Union, where the system of seed supply by specialised farms was widespread, farm-saving of seed is on the increase.

The formal sector furthermore meets difficulties serving the needs of small and resource poor farmers in marginal conditions because:

- these farmers commonly need small quantities,
- distribution needs to be arranged over wide and relatively inaccessible areas, and
- the seed demand varies strongly between years, depending on average yield levels in the foregoing production season, and on the availability of cash.

For the formal seed sector, these considerations have important implications for the quality and costs of the distributed seed, and the availability of seed at right time and right place. Underlying these factors is the relative effectiveness of the farmer's seed system.

INTEGRATED FUNCTIONING OF FARMERS' SEED SYSTEMS IN DEVELOPING COUNTRIES

On-farm selection and storage, and exchange of seed between farmers forms an integrated system. Practices of crop production, seed selection and seed storage exert selection pressures on the genetically heterogeneous local varieties. In combination with the natural selection pressure, these farmers' practices constitute a process of local crop development (Harlan, 1992; Almekinders et al., 1994). The farmer seed system is embedded in the system of crop production for household food consumption and other uses, including marketing. The harvested grains can be used for consumption, as seed for the next planting, or marketed as grain or used for seed by other farmers (Figure 1). The farmers' seed system is thus a system in which farmers produce seeds while at the same time practicing a form of crop development and maintain crop genetic diversity *in situ.*

Most often the seed is seperated from the bulk production after, at, or even before harvest, depending on the crop and farmer practices. The attention and knowledge involved in seed selection is, however, extremely variable between crops, regions, and communities, and important differences can exist between households within communities. Farmers in West Africa extract tomato seeds from the last left-over fruits that could not be marketed (Grubbe, pers. comm., see Linneman and de Bruyn, 1987); farmers in Meso America select seed from maize from the bulk of cobs after harvest or during the storage season. Al-

FIGURE 1. The multiple uses of seeds in the farmers' seed system

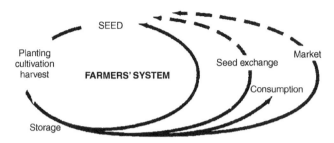

though not common, some farmers select maize cobs in the field. Selection of heads of rice or sorghum landraces or landrace mixtures in the field before harvest seems more common in Africa and Asia (Van Oosterhout, 1993; McGuire, 2000; Woldeamlak, 2001; S. Ceccarelli, pers. comm.). This way farmers ensure that the mixture of genotypes that is sown has a similar composition each season, thus eliminating largely the effect of the last growing season on the genetic composition of the population.

Women play an important role in seed selection and storage, particularly in Africa and indigenous cultures in Asia and Latin America. Consequently, they possess important knowledge about seeds and varieties. In the past decades, the importance of the distinct role of women in seed and variety management has to a large extent been overlooked or neglected. This has led in many instances to unsuccessful introductions of improved varieties or technologies. For example, criteria like cooking quality, easiness of grain processing are important characteristics of varieties, which are often not addressed in breeding programmes because only male farmers have been included in research and extension activities.

There are also cases in which the seed is produced in separate fields; often these are fields that have more favourable soil, water or disease conditions. For example, farmers in the Andes are known to plant high quality potato seed in separate fields, and avoiding mixture with the rest of their materials that may have high levels of virus infection. Off-season seed production and regional specialisation in seed production can be found in the farmers' system as well, often in combination with seed exchange between farmers. Soybean seed produced by specialised villages in East Java is a source of fresh seed for farmers in other villages with slightly different growing conditions (Linneman and Siemonsma, 1987). In Peru, farmers obtain fresh potato seed with low levels of virus infection from fields or communities located at higher altitude.

Local Crop Development

Seed selection by farmers over seasons exerts selection pressure on populations of genotypes through the criteria used by the farmers to select the seeds and through the environment (Harlan, 1992). The result of this process are landraces that developed adaptation over time to prevailing stresses such as low fertility and drought, principally because most adapted genotypes survived and contributed most seeds to the next generation. This improvement under local conditions through use and selection by farmers has in many cases also resulted in landraces that produce a harvest in the seasons with severe stresses. The yield stability that landraces acquired this way is crucial for small-scale farmers to survive in marginal environments where environmental variation and risks of crop production are high. Looking at farmers' seed production from the perspective of a plant breeder, farmers practice a form of local crop development when they produce and select seeds on farm; they maintain genetically heterogeneous varieties and their adaptation to the environmental conditions.

New genetic diversity is introduced into the farmer's system through the introduction of new varieties and introgression of genes from hybridisation with wild species or varieties. New varieties enter the farmer seed system through exchange of seeds with other farmers, seed from projects or commercial enterprises. Farmers usually evaluate newly received varieties before planting larger fields. Farmers may also pick up off-type plants, separate those seeds, and thus develop their own new variety. This is, for example, reported for rice farmers in Sierra Leone (Richards, 1987; Longley, 2000). A farmer in the Philippine selected a rice head with red grain in the field of an improved IRRI variety with improved pest resistance; the material became widely distributed under the name Bordagol (Salazar, 1992). In cross-pollinating crops, hybridisation between varieties in neighbouring fields can give rise to new gene combinations. Such new gene combinations may be selected by the farmer if it favours good plant performance. Farmers that settled in the Atlantic Coastal region of Costa Rica developed maize materials this way. These materials combine genes from local varieties they brought from their native places in Guanacaste and commercial varieties they purchased locally. Neither the varieties from Guanacaste nor the commercial varieties were adapted to the tropical lowland conditions of the Atlantic Coastal region, but over time, with hybridisation between the two types and with farmers' selection of seed, a relatively well-adapted material developed. In other situations, farmers are able to maintain varieties with their distinct characteristics through seed selection, although cross pollination between the two takes place (Almekinders and Louette, 2000; M. Fuentes, pers. comm.). In areas where wild relatives of crops still occur in the proximaty of farmers' fields, introgression from wild into cultivated species may occur. This is, for example, documented for maize in Central America, sorghum in Ethiopia, and rice in West Africa (Louette, 1999; McGuire, 2001; van Oosterhout, 1993; see also Jarvis and Hodgkin, 1999). In Ethiopia and Zimbabwe, farmers do not rogue the plants that develop from seeds containing genes from the wild species.

Seed Replacement and Off-Farm Seed Sources

Different sources of seed have different characteristics, which makes them more or less suitable, depending on the situation (Table 3). As described earlier, farmers frequently procure seed from other farmers to renew and replace their degenerated seed. Seed replacement varies per crop, it occurs probably

TABLE 3. Characteristics of seed sources and their general suitability (*) in relation to the demand for seed as planting material and as source of new varieties (adapted from Almekinders and Louwaars, 1999)

Seed sources	Characteristics	Source for planting material	Source for new varieties
On farm	Known quality, cheap, readily available	+++	− − −
Neighbours, friends & relatives (in the community)	No cash involved, readily available	++	+
Others in the community	No cash involved, readily available, not necessarily easily accessable (social differentiation)	+	++
Local market	Unreliable quality, last seed resource	− −	− −
Middle men	Non-cash arrangements/loans, unreliable quality	+, −	−, +
Neighbours, friends & relatives (outside the community)	Non-cash arrangement, resources needed for traveling	+	+++
Stores & commercial enterprises	Cash for seed and traveling	+	++
Seed agencies public seed sector	Unreliable availability and quality unknown	−	+++

(*) ranging from +++ (generally very suitable) to +, − (reasonable suitability, depending the situation) and − − − (generally unsuitable)

more frequently in the case of potato, which easily degenerates through the build-up of virus infection over generations than in wheat or rice. Another important reason for farmers to procure seed off-farm is the fact that many farmers are not able to save seed for next planting. Households often finish the grain before planting because last seasons' production was low or because urgent cash needs for schooling, medicine, funerals, etc., forced them to sell-off their harvest. In general, saving seed for next season is most difficult for the poorest farmers. The better-off farmers more frequently have sufficient surplus grain left in store at planting time to supply those farmers who could not save seed. However, the better-off and poor farmers in a community are not always socially well-connected. Cases are reported in which poor farmers do not have access to seed from richer farmers or in which poor farmers rather use consumption grains of unknown quality from the local market or improved seeds from a project than having to approach richer farmers (Sperling et al., 1993; McGuire, 2001).

For seed exchange between communities, well-developed seed exchange channels and mechanisms my exist through which farmers exchange new materials, acquire fresh seed to replace degenerated seed, and seed of varieties that were lost. In the Andes, yearly fairs between the cropping seasons had traditionally been important in the seed exchange mechanisms (Tapia and Rosa, 1993). Middle men traditionally play an important role in potato seed flows between communities in the Andes. Also seed exchange practices in other regions are reported to represent a dynamic system of supply and diffusion of seeds (Cromwell, 1990). For example, new rice varieties in Nepal, were effectively and rapidly diffused through local exchanges (Green, 1987). The national potato programme in Peru tapped into this system in 1985-1990 in order to diffuse high quality clean tuber seed of local and improved varieties (Scheidegger et al., 1989).

Poorly Connected, But Complementary Seed Systems

In most countries, the formal seed sector is not organised to support the farmers' seed system. The linkage between the two systems is usually poor, which can be partly explained by their distinctness in character. Whereas the farmer's system is an integrated system of seed production, crop development and *in situ* conservation, with seed exchange connecting the households and communities into larger seed system units, the formal sector is organised as a chain (Figure 2). In this chain, seeds and genes flow from genebanks to breeding programmes, and onwards to seed production and distribution programmes, in order to arrive as an end-product in farmers' fields to be it is used as an external input in the production of a marketable crop. In addition, this one-directional flow of the genetic diversity in a crop narrows down to a relatively small number of varieties. The collection of 'raw' local materials from farmers' fields to

FIGURE 2. The farmer and formal seed system: two parrallel functioning systems with relatively little interaction (adapted from De Boef et al., 1995)

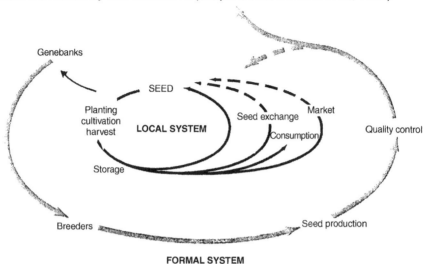

be stored in genebanks, and the distribution of end products to farmers, are the only intentional points of linkage between the two systems. The poor connection is particularly visible where these end-products of the formal system are not well-adapted to farmers' conditions. As explained above, this is particularly the case in low-input agricultural production systems in marginal areas. Farmers in these systems typically use various crop varieties, which may be either local and/or improved varieties. The first seed of improved varieties is often obtained from other farmers–and not directly from the formal sector–in other situations farmers only use local varieties because the improved varieties do not perform well in their conditions. The adaptation of the local varieties to marginal agro-ecological conditions may be represented by rare genes or gene combinations in these varieties. Therein lies a remarkable paradox that the extensive use of these 'natural resources' has not generated benefits to the farmers who initially selected and maintained theses resources (see also Visser, this volume).

Linking and Integrating the Formal and Farmers' System

In many places and crops, farmers' seed systems operate relatively independently and isolated. This indicates the relative strength of these systems in comparison to the formal system that in many cases needs the continuous support of research investments, tax holidays, etc., to supply seed at commercial and social optimum levels. Nevertheless, this does not imply that the farmers' system is without shortcomings. Farmers' seed systems are variable in performance; their effectiveness differs between crops, varieties, and farmers. Their strength and weaknesses vary strongly, and it is difficult to generalise. Some weaknesses are, however, apparent and frequent.

Need for supporting the farmers' seed system. In most cases, farmers need new, additional crop genetic diversity, despite the relative wealth of diversity in their own system. New emerging diseases create a need for resistant varieties, changing rainfall patterns may ask for earlier maturing varieties, reduced soil fertility and elimination of fertiliser subsidies require better low-input adaptation, new production technologies or new market opportunities ask for different crops or variety characteristics. The need for varieties with higher productivity is evident in practically all situations. Also seed selection practices may not be optimal and there is scope for improving the performance of local varieties and on-farm maintainance of improved varieties. Farmers often do not recognise the effects of crop pests and diseases on seed quality and crop performance next season. Storage conditions and practices result in suboptimal seed quality as well. Social or geophysical barriers between social classes and communities may reduce farmers' access to local seed sources. Traditional seed exchanges mechanisms that have eroded under the pressure of modernisation or migration may affect the resilience of farmers' systems to recover seed from lost varieties. And, finally, natural or man-made disasters may disrupt farmers' seed systems.

The expertise and technologies on several of the above-mentioned points are available in the formal system. Until now, however, this expertise has hardly been applied to strengthen the farmer's system. So far, the expertise has been used to build a formal seed sector targeting high input agriculture. For developing countries, this has in general improved crop varieties and seed supply in the more favourable production areas, but has also left a large part of the agricultural sector practically unaddressed. Apparently, for this neglected sector other approaches are needed.

Linkages in crop development. The complexity and variation of marginal environmental conditions creates great difficulties for breeding programmes to overcome genotype × environment interactions. Supplying and distributing small and variable volumes of seeds to areas that are remote and relatively inaccessible are logistically complicated and significantly increase costs of the seed. A viable strategy to increase effectiveness of the formal seed sector may be to build on the strengths of the farmer's system, and improving it where it seems to fail. Plant breeders can involve farmers in selection of varieties, identification of desirable traits, and on farm evaluation. Participatory Crop Improvement follows this approach and seeks to enrich the farmer's seed system with new genes and adapted materials (Almekinders and Elings, 2001), and increasing farmers' capacity to evaluate and select good performing plants. Improved varieties as well as local varieties from elsewhere or from stored genebank collections may be (re-)introduced via such activities.

Linkages in seed production and handling. Improving farmers' seed production and storage practices can significantly improve farmers' seed quality. Individual farmer experts may become key seed distributors in a region, being a source of improved materials and quality seed. Such seed experts may even become local (formal) seed producers (see Kugbei and Bishaw, this volume). Furthermore, training in the identification of (seed transmitted) diseases may contribute importantly to improving seed health status of on-farm produced seed.

Linkages in seed distribution. Several types of activities can contribute to imporve the flow of seed within the farmers' systems and the influx from the formal sector. Traditional seed diffusion and exchange mechanisms may be restored or enforced through the organisation of seed fairs. Community seed banks can contribute to community seed security. They can also become a central activity in a community around which community organisation and farmer training in seed production and selection can be arranged.

Strengthening of the farmer's seed system will increase seed security– which in turn enables farmers to continue to grow and maintain *in situ* the genetic diversity of their preference and choice. Such strengthening will thereby contribute simultaneously to the sustainability of farmers' livelihood and *in situ* maintenance of crop genetic diversity (Almekinders et al., 2000).

Various efforts in these directions are described in other papers in this volume. The linkage of formal institutions with the farmers' system not only serves the strengthening of the latter, but also serves the formal system to establish collaborative relationships with the farmers' system. Genebanks can thereby link their *ex situ* conservation with complementary *in situ* conservation through (re)-introducing materials lost by farmers or storing *ex situ* those materials which are losing their attractiveness for farmers. Plant breeding programmes can use farmer-participation in evaluation and selection of materials to overcome Genotype × Environment interactions. On-farm seed production may be improved with the expertise of seed technicians. And, finally, well-functioning farmers' seed systems can reduce the burden of inefficient production of seeds that offers little opportunity for privatised enterprises (bulky seeds of self-fertilising crops). A formal seed sector that can build on a functional farmers' seed systems may concentrate on producing relatively small but crucial amounts of high quality seed to be flushed into the farmers' system at adequate moments and points. NGOs can play an important intermediary role in linking the formal seed sector actors with the farmer communities. In such a way, the formal and farmer's systems may develop into an integrated seed system in which each partner has a role to play in seed production, crop development, and conservation (Figure 3).

FIGURE 3. The farmers and other organisations with multiple linkages in an integrated seed system (adapted from Almekinders et al. 2000)

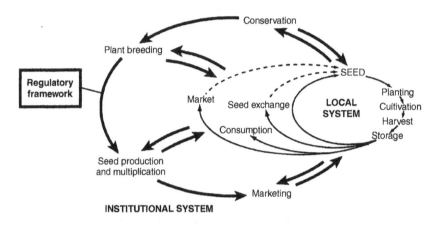

Technical support from formal seed sector specialists is essential for improving the knowledge and capacities in the farmers' system. Recognition and integration of the farmers' system in national seed policies and seed regulation is also vital to the development of linkage and integration between the formal and farmers' system.

Although the importance of farmers' seed system is mostly associated with developing countries, it is also an area of interest for economies in 'transition' and developed countries. In these countries, farmers' seed systems have been strongly discouraged for many years. Currently, however, there is a call for increased farmers' initiatives in seed production. In 'transition' countries, this is due to the fact that the formal seed supply infrastructure is unable to effectively respond to the farmers demand for seed. In the developed countries, the need for more ecological agriculture challenges the formal seed sector to supply ecologically produced seeds of niche-varieties.

REFERENCES

Admasu T. & P. Struik, 2000. Research supporting the genetic diversity of enset in southern Ethiopia. In: C.J.M. Almekinders & W.S. de Boef (eds.), *Encouraging diversity. Plant genetic resource conservation and crop development.* IT Publications, London. pp. 245-249

Almekinders, C.J.M., & A. Elings, 2001. Collaboration of farmers and breeders: participatory crop improvement in perspective. *Euphytica* 122:425-438.

Almekinders, C., W. de Boef, & J. Engels, 2000. Synthesis between crop conservation and development. In: C.J.M. Almekinders & W.S. de Boef (eds.), *Encouraging di-*

versity. Plant genetic resource conservation and crop development. IT Publications, London. pp. 330-338.

Almekinders C., & D. Louette, 2000. Examples of innovations in local seed systems in Mesoamerica. In: C.J.M. Almekinders & W.S. de Boef (eds.), *Encouraging diversity. Plant genetic resource conservation and crop development*. IT Publications, London. pp. 219-222.

Almekinders, C.J.M. & N.P. Louwaars, 1999. *Farmers' seed production. New approaches and practices*. IT publications, London. 290 p.

Almekinders, C.J.M., N.P. Louwaars & G. de Bruijn, 1994. Local seed systems and their importance for an improved seed supply in developing countries. *Euphytica* 78:207-216.

Aquino, P., F. Carrión & R. Calvo, 1999. Selected maize statistics. In: *CIMMYT 1997/98 World maize facts and trends; maize productivity in drought stressed environments: technical options and research resource allocation*. Mexico D.F., CIMMYT. pp. 41-62.

Bellon, M., 1996. The dynamics of crop infra specific diversity: a conceptual framework at the farmers' level. *Economic Botany* 50:26-39.

Bishaw, Z., & S. Kugbei, 1997. Seed supply in the WANA region–status and constraints. In: Rohrbach, D.D., Z. Bishaw, & A.J.G. van Gastel (eds.), *Alternative strategies for smallholder seed supply*. Proceedings of an International Conference on Options for Strengthening National and Regional Seed Systems in Africa and West Asia, 10-14 March 1997, Harare, ICRISAT, Patancheru, pp. 18-33

Brush, S.B., 1986. Genetic diversity and conservation in traditional farming systems. *Journal of Ethnobiology* 6:151-167.

Brouwer, J., L.K. Fussell & L. Hermann, 1993. Soil and crop growth micro-variability in the West African semi-arid tropics: a possible risk-reducing factor for subsistence farmers. *Agriculture, Ecosystems and Environment* 45:229-238.

Byerlee, D., 1996. Modern varieties productivity and sustainability. *World Development* 24:697-718.

Clawson, D.L.1985. Harvest security and intraspecific diversity in traditional tropical agriculture. *Economic Botany* 39:56-67.

Corrales, L., F. Trece Ramos & N. Uclés, 1991. *Proeycto de investigación sobre sistamas locales de provisión de semillas en Honduras*. CICTEC/IVO, Tegucigalpa/Tilburg.

Cromwell, E., E. Friis Hansen, & M. Turner, 1992. *The seed sector in developing countries: a framework for performance analysis*. Working Paper 65, ODI, London, UK.

Cromwell, E. (ed.), 1990. *Diffusion mechanisms in small farm communities: lessons from Asia, Africa and Latin America*. Network paper 21, ODI, London, UK.

DANAGRO, 1988. SADCC *Tregional seed prodution and supply project: main report (vol. 1A) and coutnry reports (vol. 2A-2J)*. DANAGRO Adviser, Glostrup, Denmark.

Douglas, J.E., 1980. *Successful seed programs. A planning and management guide*. Westview Press, Boulder, CO.

FAO, 1996. *Global plan of action for the conservation and sustainable utilization of plant genetic resources for food and agriculture*. Food and Agriculture Organisation of the United Nations, Rome.

FAO, 1994. *FAO seed review 1989-1990.* Food and Agriculture Organisation of the United Nations, Rome.

Ghijsen, H., 1996. The development of variety testing and breeders' rights in the Netherlands. In: H. v. Amstel, J.W.T. Bottema, M. Sidik, & C.E. van Santen (eds.), *Integrating seed systems for annual food crops.* Proceedings of a workshop held in Malang, Indonesia, October 24-27, 1995. CGPRT Centre, Bogor, pp. 223-225.

Green, T., 1987. *Farmer-to-farmer seed exchange in the eastern hills of Nepal: the case of 'Pokhreli masino' rice.* Kathmandu, Nepal, Pakhribas Agricultural Centre, working paper 05/87

Harlan, J.R., 1992. *Crops and man.* 2nd ed. Am. Soc. of Agronomy & Crop Science Soc. of America, Madison, WI.

Hernández, E., 1985. Maize and man in the Southwest. *Economic Botany* 39(4): 416-430.

Jarvis, D., & T. Hodgkin, 1999. Wild relatives and crop cultivars: detecting natural introgression and farmer selection of new genetic combinations in agroecosystems. *Molecular Ecology* 8:159-173.

Kutay, A., 1997. The private seed sector in Turkey. In: Rohrbach, D.D., Z. Bishaw, & A.J.G. van Gastel (eds.), *Alternative strategies for smallholder seed supply: Proceedings of an International Conference on Options for strengthening national and regional seed systems in Africa and West Asia,* 10-14 March 1997, Harare, Zimbabwe, ICRISAT. pp. 49-53.

Kerr, J. & S. Kolavalli, 1999. *Impact of agricultural research on poverty alleviation: conceptual framework with illustrations from literature.* EPTD Discussion Paper, IFPRI, Washington. 195 pp.

Kloppenburg, J.R.,1988. *First the seed. The political economy of plant biotechnology 1492-2000.* New York, Cambridge Unversity Press.

Linnemann A. & J. Simonsma, 1987. Variety choice and seed supply by smallholders. *ILEIA Newsletter* 5(4):22-23.

Linnemann, A.R., & G.H. de Bruyn, 1987. Traditional supply for food crops. *ILEIA Newsletter* 3(2):10-11

Louette, D., 1999. Traditional management of seed and genetic diversity: what is a landrace? In: S.B. Brush (ed.), *Genes in the field. On-farm conservation of crop diversity.* IPGRI/IDRC/Lewis Publishers, Boca Raton. pp. 109-142.

Lipton, M. & R. Longhurst, 1989. *New seeds and poor people.* Johns Hopkins University Press, Baltimore, MD.

Longley, C.A., 2000. *A social life of seeds: Local management of crop variability in North-Western Sierra Leone.* Dept. of Anthroplogy, University of London. PhD Thesis. 306 p.

Louwaars N.P. & G.A.M. van Marrewijk, 1996. *Seed supply systems in developing countries.* CTA, Wageningen, The Netherlands.

McGuire, S., 2001. Farmers' management of sorghum genetic resources in Ethiopia: a basis for participatory plant breeding? In: D. Cleveland and D. Soleri (eds.) *Collaborative plant breeding: Integrating farmers' and plant breeders' knowledge and practice.* Wallingford, Oxon: CABI.

NN, 1999. *Seed Sector Programme,* Ministry of Agriculture and Land Reclamation, Agricultural Services Sector/Deutsche Gesellschaft für Technische Zusammenarbeit

(GTZ) GmbH. Cooperation between Arab Republic of Egypt–Federal Republic of Germany.

Prain, G., J. Schneider, & C. Widiyastuti, 2000. Farmers' maintenance of sweet potato diversity in Irian Jaya. In: C.J.M. Almekinders & W.S. de Boef (eds.), *Encouraging diversity. Plant genetic resource conservation and crop development.* IT Publications, London. pp. 54-59.

Richards, P., 1987. Spreading risks across slopes: diversified rice production in central Sierra Leone. *ILEIA Newsletter* 3(2):8-9.

Salazar, R., 1992. MASIPAG: alternative communtiy rice-breeding in the Philippines. *Appropriate Technology* 18(4):20-21.

Scheidegger, U., G. Prain, F. Ezeta, & C. Vittorelli, 1989. *Linking formal R & D to indigenous systems: A user oriented potato seed programme for Peru.* Agricultural Administration (Research and Extension) Network Paper 10. ODI, London.

Sperling, L., M.E. Loevinsohn & B. Ntambovura, 1993. Rethinking the farmer's role in plant breeding: local bean experts and on-station selection in Rwanda. *Expl. Agr.* 29: 509-519.

Tapia, M.E. & A. Rosa, 1993. Seed fairs in the Andes: a strategy for local conservation of plant genetic resources. In: W. de Boef, K. Amanor, K. Wellard & A. Bebbington (eds.), *Cultivating knowledge. Genetic diversity, farmer experimentation and crop research*, pp. 111-118. IT Publications Ltd., London.

van Oosterhout, S., 1993. Sorghum genetic resources of small-scale farmers in Zimbabwe. In: W. de Boef, K. Amanor, K. Wellard & A. Bebbington (eds.), *Cultivating knowledge. Genetic diversity, farmer experimentation and crop research.* IT Publications Ltd., London. pp. 89-95.

Woldeamlak, A., 2001. *Mixed cropping of barley (Hordeum vulgare) and wheat (Triticum aestivum) landraces in the central highlands of Eritrea.* PhD Thesis, Wageningen University.

Seed Sectors in Transition:
From Centrally Planned to Free Market

Gary Reusché

SUMMARY. The seed sectors in former Soviet countries that have developed under a centrally planned economy need to respond to the changing business environment following perestroika. In the Soviet times, each step in the process from breeding through multiplication and quality control, and distribution had its own institutions (public) with their assigned task, group of professionals, and in theory all the pieces fit together into a complete system. The opening of the economy caused a severe break-up of the research base chasing short-term financial gains, a break down of investments in seed production infrastructure due to hasty privatisation of farms and facilities, and poorly developed markets.

Strong bureaucratic procedures, slowly changing institutions and regulatory systems, and still inefficient agriculture with low purchasing power are major challenges for the emerging seed industry. The development of an effective seed sector in these countries will depend on the creation of transparent seed policies, regulations and public institutions that support the sector. It will also need a number of vital changes outside the seed sector itself, such as a restructuring of the banking sector, both for credit facilities to the seed suppliers and for the agricultural sector at large in order to create buyers. Transformation is, however, primarily a "change of beliefs and assumptions" to achieve a new commitment, motivation, empowerment, emotional intelligence, innovation and creativ-

Gary Reusché is Consultant, Moscow, Russian Federation.
Address correspondence to: Gary Reusché, Ultisa Mishina 28/32-25, 125083 Moscow, Russia.

[Haworth co-indexing entry note]: "Seed Sectors in Transition: From Centrally Planned to Free Market." Reusché, Gary. Co-published simultaneously in *Journal of New Seeds* (Food Products Press, an imprint of The Haworth Press, Inc.) Vol. 4, No. 1/2, 2002, pp. 35-45; and: *Seed Policy, Legislation and Law: Widening a Narrow Focus* (ed: Niels P. Louwaars) Food Products Press, an imprint of The Haworth Press, Inc., 2002, pp. 35-45. Single or multiple copies of this article are available for a fee from The Haworth Document Delivery Service [1-800-HAWORTH 9:00 a.m. - 5:00 p.m. (EST). E-mail address: getinfo@haworthpressinc.com].

ity. The emerging new generation will be able to build up a new system of efficient seed enterprises in Russia. *[Article copies available for a fee from The Haworth Document Delivery Service: 1-800-HAWORTH. E-mail address: <getinfo@haworthpressinc.com> Website: <http://www.HaworthPress. com> © 2002 by The Haworth Press, Inc. All rights reserved.]*

KEYWORDS. Seed enterprise, transformation, plan-economy, seed policy

SOVIET LEGACY

Inefficient Production Base

The transition from a centrally planned economy to a market economy in Russia, and similarly throughout much of the ex-Soviet Union, thoroughly disrupted the Soviet-era seed industry, and in many cases destroyed its basis altogether. But even this characterisation is not sufficient to describe the situation of the seed industry in the Soviet Union immediately following perestroika and during much of the 1990s. To this has to be added the fact that Soviet-era agriculture was anecdotally inefficient due to widely differing circumstances:

  i. the forced collectivisation and Stalin-era purges of farmers (kulagi and peasants) and their replacement by workers reduced commitment at the farms,
  ii. wide differences in funding for various agricultural collectives depending apparently on the closeness of the connection to the communist hierarchy, and
  iii. more than a decade of underfunding due to the same processes that caused the break up of the Soviet state.

From the perspective of Western European or North American seed industry specialists, the Soviet organization of the seed industry ran counter to the trends that shaped the seed industries in other countries of the developed world. In North America and Europe the early state roles of the state in shaping breeding and seed production were gradually and naturally privatized for the more commercial aspects of the industry. The entire seed industry in the ex-Soviet Union on the other hand was fully in the state sector until the time of perestroika. No one was ready for the catastrophic events that took place in the seed sector after the break-up of the Soviet Union.

The centrally directed Soviet seed system was, outwardly, a symmetrically designed system of research institutes, institutes for seed inspection and variety testing, seed processing facilities, basic seed farms, farms for the multiplication of subsequent generations, and distribution networks often based in regional and local agricultural departments. Each step in this process had its assigned task, group of professionals, and in theory all the pieces fit together into a complete system. In practice, however, similar to other centrally directed sectors of the economy, the lack of local incentives, initiatives and decision-making authority frequently made the realization of plans something less than desired.

The system was designed to make the Soviet Union self-sufficient in seeds, but it frequently required imports of seeds, particularly vegetables and potatoes. There were few parallel markets, with the exception of a wide-spread country home culture of saving vegetable, flower and potato seeds and sharing these with friends and relatives.

In addition to the above, there were parts of the system that appeared to originate from autocratic decision-making. For example, the hybrid maize industry was based in the same organization as the grain storage organization (that managed the large grain elevators). Western experts visiting these facilities in the early 90s were told that they were built on the "American model" after the visit of Khrushev to the mid-west of the United States. However, when this technology was transplanted in the Russian system of the grain elevators, a bizarre system of horizontal conveyors connected widespread buildings due to the explosion hazard of grain dust in the processing facilities, indirectly exposing the mentality of designers of the system and their orientation towards grain elevators.

Break Up of Research Base

No seed industry can exist without a research base, and comprehensive variety development and testing programs. Research in the centrally directed economy was undertaken entirely at regional and sub-regional research institutes located in various geographic regions of the Soviet Union. At the time of perestroika, not only did these institutions loose much of their source of funding due to the economic crises, but the institutions that were originally planned for the benefit of the Soviet Union often ended up in completely different independent countries. Thus, parts of the research required for the Russian Federation, for example, may have been located in Kazakhstan or Ukraine. Often the staff got demoralized due to lack of investments in facilities, and simply because they were often not paid. The research institutions lost contact with international developments such as the use of modern information technology and biotechnology in breeding. Schemes were devised to develop income-producing activities, even where such activities were against the administrative statues of the institution. By the end of the 90s, many of the more progressive institutes had begun to utilize their genetic and program resources and take on more and more commercial activities and supply seeds for cash. Smaller research institutions fully transformed into seed production units while totally disregarding any further research that is considered un-productive (at the short run).

Outdated and Decaying Infrastructure

Nobody visiting the ex-Soviet Union can be unmoved by the situation in the rural areas. Two outstanding features are mentioned. First, the sector appears chronically under-funded. With the exception of a very few special collective farms which were run by local communist party leaders, the majority of the collective farms are run-down, the equipments are old, its design is outdated, and its buildings and administrative structures are of very basic construction, often lacking such luxuries as indoor plumbing and toilets. This situation affects not only the capacity to produce seed, but also the potential profitability of the agricultural sector in general, and thus the ability of the farms to purchase professionally grown seed. By the turn of the millennium, the permanence of the economic changes in the countries had been accepted, but the largest majority of the large collective farms are technically bankrupt. They have been stripping assets for more than 10 years, have little or no access to seasonal credit, and are often the victims of an imperfect and oftentimes monopolistic commodity marketing structure.

Problems Created by the Privatization Process

The privatization of the agricultural sector in the ex-Soviet Union took place overnight. In the beginning, no one believed it was really privatized, as the old state structures still dominated the sector. However, during the 90s, the state structures no longer had the funding to dominate, and subsidies for the agricultural sector continually declined, leaving each individual enterprise in the sector responsible for its own success and failure for the first time. Most agricultural sector enterprises, from collective farms to agro-processing facilities, were handed over to the workers of the enterprises, including retired workers. Thus the workers became the owners, and business decisions were complicated and hard decisions were difficult to make. In the Soviet system, many social services were not delivered by local governments, but by the enterprises themselves, and thus the newly privatized enterprises were required to pay for these social services from their commercial revenues. During the 1990s, many of these services have theoretically been transferred to local government, but the process proceeds very gradually. At the beginning of the 90s, the privatized commodity distribution companies were large, centralized, and too few in number to create a competitive situation. As initially the farms had nowhere else to sell their commodities, and nowhere to store them, they were initially forced to sell to these companies even when it meant taking loses in an inflationary environment. During the 90s, new marketing outlets were developed, and the lorries of various farms in the regions were often found marketing their commodities directly to consumers. Finally, almost the entire agricultural sector during the Soviet area was focused on production, and meeting production targets. After privatization, they had to become more conscious of cost-benefit relations, and create marketing structures. In most cases, the staff was not trained for these functions.

REBUILDING

World Bank Involvement: Modernization of Seed Production

Given the above characterization, the justification for the involvement of the World Bank and other multilateral and bilateral development activities in the seed sector is clear. Given the size of the ex-Soviet Union and Warsaw Pact countries, it is impossible to describe the on-going process of rebuilding in its entirety. The following discussion is based on World Bank projects in Russia and the Ukraine, and some personal contacts with Western seed companies wishing to do business in the region.

Vertically Integrated

Rebuilding of the seed industry often focused on the seed enterprise. "Western" models of vertical integration, including R&D, internal quality control, organization of production from basic to commercial seed, demonstrations, and marketing were considered not only appropriate, but perhaps the only way to redevelop the industry.

Modern Design

In order to sell high quality seed, and assure the repayment of credit, it was considered necessary to support the investment in new equipment and facilities, including selected support to seed production, as well as modern seed processing facilities. For example, the existing maize seed pickers did not shuck the cob. Maize seed farms literally dumped huge piles of maize, and hundreds of women were employed to shuck the maize, before it was sent for drying. This step created a huge bottleneck, and reduced seed quality. The development activities focused primarily on the high-value seed sector, initially hybrid maize and sunflower, vegetables and potatoes. Feasibility studies, technical and financial, were positive and investment financing was obtained for a number of enterprises.

Modern Management

Modern management encompasses a range of topics, but in this context the focus was on appropriate staffing and management structure, management information systems, internal quality control, marketing, and financial control.

Harmonization of Regulations

In addition to activities focused on the enterprises, other EU and bilateral projects supported modernization of the variety testing and registration organization, and the seed certification and regulatory agency. In many cases, the staffing of these structures included many talented and capable professionals. However, adjustments in their operations and financing, and new rules and regulations had to be made to adapt to a free market environment and increased interaction with other countries on seed related matters. Among the changes implemented was the adoption of UPOV inspired legislation providing breeders with property rights to their varieties, promising the financial basis for investments in breeding in the private sector. As little investments are being made by the government sector, this was a key decision.

International Trade: Import and Sell

The activities of most foreign seed companies have been primarily to register, import and then market their seeds in the newly emerging markets of the ex-Soviet Union. Few foreign seed companies have chosen to invest in the region, either for local production, or for R&D (except for variety testing). Examples can be cited where companies with adapted, self-pollinated species do not want their materials to be sold in the Russian Federation for fear of non-authorized reproduction and sales in the imperfectly established regulatory environment in much of the ex-Soviet Union. The huge potential market in the region is conservatively approached due to the risks of doing business, or simply the difficulty to understand how best to approach an apparently chaotic marketplace with massive problems. Many companies look at investment possibilities, even at a time of risk, as if operations can be established at this time, and markets achieved, when times are better they would be in a very favorable position to grow quickly. However, it would appear that the short-term risks and uncertainties outweigh the possible long-term benefits at this time.

SEED ENTERPRISE DEVELOPMENT

The initial development of seed enterprises after perestroika was based on privatized structures that existed at the end of the communist period. Foreign investors visited the region, and reflected on getting started and capturing new markets, but few did anything more than attempt to export and sell seeds. Most seed companies understood that such activity was short-sighted, but nonetheless the risks of operating in Russia during the 1990s tended to overshadow any ideas of capturing new markets.

As a seed company is a small and medium enterprise (SME), it is impacted by the business environment like any SME. Legal small entrepreneurship in Russia (and the ex-Soviet Union) was literally generated by perestroika. The date of its birth is 18 July 1991 when a government decree set the criteria for a small enterprise in Russia and specified the general terms and rules of their operation. During the last 10 years the legislative basis, which regulates entrepreneurial activity, particularly, small business, has been created. The goals and objectives of the governmental policy towards small business support have been identified. Mechanisms for implementing the support measures have been developed; organizations, which can perform them, have been created. A network of business support organizations, which provide assistance to SMEs in training, information support, consulting, and financial assistance has been established. This context is necessary to understand why things moved so slowly during this period. Things are far from perfect, but the macro-environment has improved to a point where serious opportunities now exist.

A number of remaining problems exist for the establishment and development of modern seed enterprises including:

- the banking sector;
- products and markets;
- reengineering and transformation; and
- regulation, corruption and ethics.

Banking Sector

At the turn of the millennium, there were nearly 1,300 banks in Russia, most of which were very small and financially distressed. Net capital across the banking system at end-1998 was thought to be less than zero on an IAS basis and was little changed as the 1990s ended. The August 1998 banking crisis led to an extended banking crisis. The state's response to the collapse of the banking system was ineffectual, and marked by political infighting, conflicts of interest, and neglect of the underlying problems. Since that time, the banking system has basically been restructuring itself.

In the weeks after the August crisis, there were enough transfers of retail deposits to Sberbank to effectively re-nationalize the majority of the retail deposit market. Systemic risk still dominates in the form of interference (leading to corruption) by federal and local authorities, economic instability, poor quality financial data, economically unsustainable taxation, and unreliable law enforcement. Individual banks are impacted both directly and because the systemic risk reduces the number and quality of potential borrowers.

Thus, agricultural firms that need investment credits and working capital are faced with, from a Western European perspective, a non-functioning banking system. Investment credits are basically not available, or, if available, only for relatively short periods of time (1-2 years). Only the largest and most successful agricultural enterprises, with strong cash flow and close relations with a well-connected bank, can obtain working capital. This situation reduces the opportunities for seed enterprises to get started from local Russian sources, but creates opportunities for foreign investors willing to bring investment credits to Russia, and provide additional credibility for a joint venture to gain a privileged position with a Russian bank for working capital.

In 2001, in various regions in Russia, the situation involving credit and investments in agriculture has begun to look better. In the middle and lower Volga regions, a new subsidiary of Sberbank, Agroprominvest, appears to be establishing itself to provide banking services to agribusiness, emphasizing chain development. This or some other similar initiative will be necessary for strong seed industry development in Russia. The amount of time required for the development of the banking sector is unpredictable, but the requirements for investment in the seed sector are everywhere obvious.

Products and Markets

Seed enterprises need products to sell and need buyers. Activities in varietal research are difficult to characterize due to regional differences and poor distribution of information. Visits to regional research institutes that had strong breeding programs during the communist era, are often depressing to Western scientists, as the facilities are not properly maintained and may appear mostly empty. Many good scientists have been lost, and research budgets are very low. These facts would strongly imply that most of the work being done is largely grounded in the past system, and new state-of-the-art breeding has yet to be developed. Some research establishments engage in seed production and use this activity to generate income, suggesting a reengineering of these institutions into companies that have many characteristics of seed companies.

The competitiveness of Russian produced varieties in such circumstances can be hypothesized to be low, and declining. A strong seed system is based on strong research programs. In most of the small grain crops, varieties typically have a life span of less than 10 years before they are replaced by new varieties, most appropriate to the disease and production environment. Russian potato varieties cannot match the yield potential of Dutch varieties, for example, but whether this is due to poor seed quality or genetic inferiority is debated by some sources. Specialized potato varieties that provide the right characteristics for fast food chains do not exist. Malting barley varieties restrict the Russian supply of malting barley for the Russian breweries, forcing a fast growing market to depend entirely on imported grains.

From another perspective, until the Russian research and development is rebuilt, there might appear to be a large opportunity for Western European seed companies to move into Russia, gain market acceptance and capture new markets. This could be beneficial to Russian agriculture, and Western seed companies. This also is not taking place, apparently due to the business environment, as well as the potential loss of control of a variety due to the non-enforcement of plant variety production regulations.

Rebuilding the Russian varietal research programs to be as effective as well funded and highly sophisticated research in Western Europe is not likely to take place overnight. The more likely scenario is that seed enterprises will begin R&D, initially on a limited basis, and based upon the returns to investments in research, this activity would develop. During this time, it may be in the best interest of Russian agriculture to improve the variety testing and registration activities, and regulation and enforcement of plant variety protection legislation, and in this way gain access to significant germplasm resources from existing multinational seed companies that may have good materials for a number of regions in Russia.

The other side of the issue, after having a good product, is being able to sell the product and make profits. Russian agriculture is still struggling, in general, to find itself in the new economy. But the situation is gradually improving, and even though it may not be statistically provable due to the extent of the unofficial economy, experienced observers can see differences in the countryside as a new group of managers on the large commercial farms combine whatever resources they have with various business and barter constructions to keep their enterprises alive and find ways to survive. These new managers are much more entrepreneurial than their communist-era predecessors.

Reengineering and Transformation

Reengineering, or the fundamental rethinking and radical redesign of business processes to achieve dramatic improvements in critical measures of performance such as quality, speed, efficiency and productivity, has been a major activity in agriculture in the ex-Soviet Union since perestroika. Large collective farms and institutes that did not choose to pursue this activity have stagnated and fallen into technical bankruptcy, its new "owners" (becoming owners during the privatization process) falling into a subsistence pattern of production. The policy of the Russian government has been constant throughout the period: "The large collective farms, as privatized and commercial entities, should not be broken up. They should work as modern commercial firms." Transition from the centrally directed collective farm to modern commercial firms managing its own assets, has not been easy.

Transformation or a "change of beliefs and assumptions" is very important in the rural sectors of the ex-Soviet Union. These sectors are often characterized as places where no one would choose to live, and the home of poverty and alcoholism. Transformation in a business environment is classically related to the characteristics and style of the management. The transformation that is required in Russian agriculture is to reintroduce elements of "belief" that were systematically de-emphasized in Soviet agriculture, and a radical redesign of its organizational culture to achieve dramatic improvements in critical measures of performance such as commitment, motivation, empowerment, emotional intelligence, innovation and creativity.

Thus, like other SMEs and agricultural firms in Russia, the seed industry will have to deal with issues of reengineering and transformation. The combination of Western investors and organizational systems, with progressive Russian experts hungry for meaningful work at a decent salary, is a model that has worked well in many sectors of the Russian economy.

Regulation, Corruption and Ethics

No discussion of seed industry trends in the ex-Soviet Union would be complete without mentioning regulation, corruption and the necessary business ethics that will make a seed enterprise successful. Many business consultants in Russia would suggest that the bureaucratic environment, predisposed to corruption due to the unreasonably low salaries of bureaucrats, is the biggest threats to the development of any business, but especially one as logistically complex as the seed industry (not lawlessness). New laws are being written, but sometimes the regulations that permit their enforcement are missing or poorly stated. Or, regulatory agencies often appear to selectively enforce regulations, or it takes far too long (adding costs) for routine regulatory processes to reach a conclusion. There is much discussion amongst the various administrative levels, including President Putin that this situation has to change fast.

It would not simply be enough to design an enabling business environment. Questions of ethics and values and business culture have also to be addressed. Young people in Russia are now growing up without first hand knowledge of communism. They are knowledgeable about free market systems, and desire to work for the normal goals of home, family, and a good standard of living. Seed enterprises in Russia can tap and inspire the new generation, inspire and motivate them with good values and ethics, and build up the new system. Experience over the last decade has shown that Russians can adapt to completely new situations with amazing speed and facility. The technical issues involved with a seed enterprise are not the main bottlenecks. The overall business environment for the SME and agricultural sector on one hand, and introducing a business culture that gains a reputation for quality, reliability, and value on the other, are the main challenges. The current situation will not remain static and underdeveloped.

Policy Measures
for Stimulating Indigenous Seed Enterprises

Sam Kugbei

Zewdie Bishaw

SUMMARY. Considering the limited success of the public sector in delivering seed to small farmers in remote areas, and the lack of commercial interest on the part of large seed companies, it seems that small and location-specific enterprises may be the best option for fulfilling this role. However, the strategy in promoting such enterprises should differ from the usual 'top-down' measures characteristic of large seed projects and companies because the conditions are different in terms of crops, resources and potential profitability.

Small-scale enterprises require community-based interventions, since these businesses are meant to serve small farmers in rural areas. Attracting investment in this form of seed supply, and creating sufficient interest in the community for the seed the enterprises produce, are formidable challenges. They need a strong commitment on the part of governments in introducing favorable policies and providing adequate incentives that can encourage investment.

The enterprises should be allowed to evolve using the community as a basis. Therefore, policy interventions should be consistent and provide adequate support and protection for both the producers and the seed using customers. Pioneer enterprises should operate without external support, such as direct subsidies even though subsidised services may be

Sam Kugbei and Zewdie Bishaw are affiliated with the ICARDA Seed Unit, Aleppo, Syria.

Address correspondence to: Sam Kugbei or Zewdie Bishaw, International Centre for Agricultural Research for the Dry Areas, P.O. Box 5466, Aleppo, Syria.

[Haworth co-indexing entry note]: "Policy Measures for Stimulating Indigenous Seed Enterprises." Kugbei, Sam, and Zewdie Bishaw. Co-published simultaneously in *Journal of New Seeds* (Food Products Press, an imprint of The Haworth Press, Inc.) Vol. 4, No. 1/2, 2002, pp. 47-63; and: *Seed Policy, Legislation and Law: Widening a Narrow Focus* (ed: Niels P. Louwaars) Food Products Press, an imprint of The Haworth Press, Inc., 2002, pp. 47-63. Single or multiple copies of this article are available for a fee from The Haworth Document Delivery Service [1-800-HAWORTH 9:00 a.m. - 5:00 p.m. (EST). E-mail address: getinfo@haworthpressinc.com].

necessary to make the environment favorable. Once competition has developed, the production and use of seed will then become sustainable since forces within the farming community will drive both supply and demand. Successful small enterprises may join forces in form of association as a means of protecting their interests and serving as a forum for sharing experiences. *[Article copies available for a fee from The Haworth Document Delivery Service: 1-800-HAWORTH. E-mail address: <getinfo@ haworthpressinc.com> Website: <http://www.HaworthPress.com> © 2002 by The Haworth Press, Inc. All rights reserved.]*

KEYWORDS. Small enterprise development, community based enterprise, local seed supply, participatory plant breeding

INTRODUCTION

National seed industries and their environments are heterogeneous in terms of crops and varieties, agro-ecology, farmers, local culture and traditions, and markets. The industries constitute enterprises with varying ownership patterns and exist in different forms and scale. The efficiency and effectiveness of a seed system depends on the extent to which its enterprises generate and use quality seed of new or existing varieties in improving productivity and welfare of the farming community. Governments have often intervened in a top-down manner as major producers and distributors of seed in developing countries. This achieved limited success, as few farmers, mostly the larger ones, have tended to use seed coming from government organizations, while the majority of small farmers continue to save their own seed of traditional and seldom use new varieties. This pattern is predominant particularly in remote areas and less favorable environments and for minor crops which are important for the livelihood and food security of small farmers.

The weakness of government supply systems has led to the adoption of policies aimed at encouraging a greater involvement of the private sector in seed supply. However, while some crops are inherently more suitable for commercialization especially those for which hybrids exist, others are less so (e.g., self-pollinating cereals and grain legumes) although these may be the major sources of food supply. Privatization has therefore been seen as concentrating on high-value crops, as often demonstrated by the activities of multinational companies and their affiliates. For many developing countries, what is needed is a privatization policy that offers a means of diversifying the sources of supply and encouraging different types of enterprises to participate profitably in the seed delivery business, while helping to attain national food self-sufficiency and security.

The varying social, cultural and economic dimensions of farming systems make the process of privatization rather complex. It is, therefore, necessary to redefine the appropriate roles for the government and other parties involved in the seed sector, such as private entrepreneurs and development agencies including Non-Governmental Organizations (NGOs). Promoting enterprise development needs a proper understanding of existing conditions particularly local institutional arrangements, farming systems and markets. Top-down interventions that are characteristic of the formal sector would hardly work in such situations, since the flow of varieties and seed is driven primarily by what happens at the local level. Instead, bottom-up approaches appear to be more appropriate because they have the potential of building upon existing knowledge, skills and experience. This is particularly relevant if the objective is to stimulate enterprises within the farming community to serve local markets. Creating an enabling environment for enterprise development requires appropriate policies that are directed at improving the economic, social, infrastructure, regulatory and legal basis for business activity.

NEED FOR ALTERNATIVE SEED DELIVERY SYSTEMS

Since the early 1970s, seed industry growth in developing countries has followed two main directions. First, was the establishment of parastatal organizations to deliver quality seed of improved varieties. This was followed by the formation of formal private seed companies, both foreign and domestic, as a consequence of structural adjustment efforts. None of these have fulfilled the variety and seed needs of a majority of farmers, mostly small-scale, who continue to produce and keep their own seed of mainly traditional varieties.

In general, the parastatal companies are large-scale and capital-intensive organizations, whose operations are heavily subsidized by national governments and foreign donors. The primary aim is to ensure that farmers nationwide have access to quality seed of new improved varieties that could serve as the basis for increasing the productivity of crops. Seed production and distribution are formally organized, starting with breeder seed of officially released varieties, through intermediate stages of multiplication until certified seed is obtained, which is the final product that is sold to farmers. Unfortunately, a majority of farmers, particularly those located in remote areas, have not adopted these improved varieties and seeds for several reasons including absence of adaptable varieties, lack of access, high prices of seed and accompanying inputs, limited awareness about the attributes of the new varieties. Perhaps most important is farmers not perceiving a significant difference between certified seed and what they already have. After several years, parastatals have not proved to be effective means of delivering seed particularly to small farmers, and many of these have encountered financial difficulties or continue to rely on subsidies (Turner, Kugbei and Bishaw, 2000).

Structural adjustment efforts of market liberalization and privatization have meant restructuring of the parastatals. There are difficulties associated with this. First, is that parastatals are generally not attractive business entities for the private sector to buy and manage profitably. Many deal in low-margin crops; they are over-staffed, and have poor facilities. Outright sale is even more difficult if such assets are established on state land. There is also the dilemma of governments wanting parastatals to operate profitably and at the same time expecting them to meet social obligations such as providing seed of less-profitable crops in which the private sector shows little interest. To be cost effective, public enterprises would wish to concentrate on hybrids, but may do so at the expense of other important but less profitable crops.

Despite their commercial limitations, parastatals have generally provided the basis for private sector development in many countries. In particular, staff and growers of parastals acquire relevant technical training and experience that make them successful entrepreneurs in private seed production. Part of this success arises from the relationships they have maintained with farmers in the seed-using community. In addition, distribution and sales networks established by parastatals often become valuable start-up assets for emerging private businesses. Turner (1999) refers to capital investments in key components of the seed program and complementary training as major developmental strengths of the early government projects.

On the other hand some countries put in place favorable policy environment and legislation including appropriate investment and financial laws, banking and credit facilities and various business incentives to stimulate the private sector and to attract both foreign and domestic capital into the seed sector. As a result, many indigenous and foreign companies started operating in the seed sector through direct investment or joint ventures, for example in countries such as Egypt, Morocco, Pakistan and Turkey (Bishaw and Kugbei, 1997). However, the private sector tends to maximize profit by focusing on hybrids and cash crops, and devotes little effort to self-pollinated crops, since these will bring little financial gain.

Limited 'flexibility' on the part of the public sector and the 'selectivity' of private sector companies render both less effective seed suppliers for the diverse markets in small-scale farming communities. Neither the public nor the private sector can handle small quantities of seed from a large number of varieties and supply these to scattered locations in remote areas.

The nature and pattern of variety and seed use amongst the majority of farmers, therefore, require new approaches to variety development and promotion, as well as seed production and marketing. The traditional crops are usually self-pollinating, making it possible for farmers to easily produce and save their own seed and maintain reasonably good quality standards as they have done for centuries. The seed they produce is a close substitute to any that may come from external sources. A new system that introduces quality seed or new varieties must be capable of competing with this existing system. As much as farmers prefer their own local materials, productivity of these are generally low and they can hardly meet the current food security needs of the rapidly growing rural population. Furthermore, changing climatic patterns now require new varieties as older ones become less adaptable. For instance, long duration varieties may no longer be suitable in places where the rainy season has become relatively short and erratic. Also, new varieties need to be developed to meet location-specific needs in cases where there is a wide diversity in agro-ecological conditions.

All these factors call for new approaches to seed delivery, particularly for small farmers. However, new institutional arrangement for seed supply should be cost-effective and technically efficient at the local level. Essential features include low overhead costs that could enable seed to be sold at reasonable prices, higher seed quality than material that farmers keep, and close involvement by the farmers themselves as both producers and users. Since large organizations have not been able to effectively fulfill these requirements, it seems more likely that small businesses could do better in such situations as they operate on a smaller scale and could therefore be better suited to location-specific conditions. In other words, this approach would mean small enterprises catering for small farmers. This must be an indigenous activity and should be based on a proper understanding of the local seed market and institutions that will determine what seed to produce, how to produce it, in what quantity, which quality standards to aim at, how to sell the final product and at which price that can generate sufficient margins for the producer.

Figure 1 illustrates a 'bottom-up' concept of seed enterprise development. The fundamental basis of this approach lies within the farming community with its local resources, indigenous knowledge and practices. The first step in formulating any measure of intervention is to fully understand how this local system functions, and then to assess its strengths and weaknesses. New varieties are normally needed to enhance on-farm productivity, but the attributes of such varieties will depend on the limitations of existing ones and the incremental improvement that is required. New varieties can only find a place in local systems when there is an effective demand for them as expressed by the genuine desire of farmers. For this to happen, farmers have to become aware of the existence of better varieties and from where they can obtain quality seed. The extension service has an important role to play on the demand side through variety demonstrations, field days and farmers' meetings. However, an even greater difficulty on the supply side is finding an appropriate means of meeting the variety and seed needs of such farmers on a sustainable basis.

FIGURE 1. Bottom-up concept of small seed enterprise development

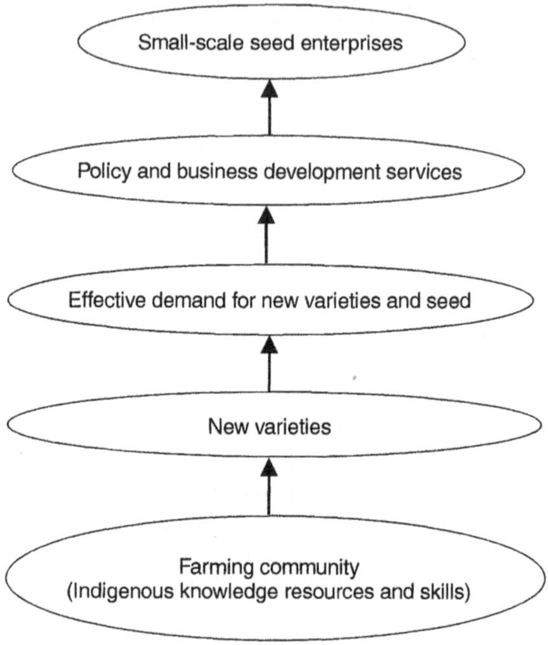

TYPES OF SMALL SCALE SEED ENTERPRISES

TYPES OF SMALL SCALE SEED ENTERPRISES

In the context of seed delivery, a small-scale enterprise is a business that is owned and managed by either one person or few people, who are engaged in not only production, but marketing of seed as well. At the community level, these may be individual farmers, a group of farmers, traders or merchants, co-operatives, farmers' organizations or associations, etc. (Kugbei, 2000). Non-governmental organizations are often active in promoting such institutional arrangements at the local level.

Individuals and Groups of Farmers

There are key farmers in any community that are known as the source of good seed, from whom other farmers obtain their supplies when they are in need. Because of the close relationship between these producers and the seed-using farmers, the business transactions they undertake are in different forms including cash purchase and exchange-in-kind. Payment may some-times be deferred until harvest time, since there is a trustworthy relationship between the supplier and the user of seed, who are normally within close proximity of one another. Farmers who could organize themselves into small groups could pool resources together, have access to a larger market and benefit from economies of scale.

Traders and Merchants

Traders and merchants located in farming communities usually deal in a range of products including tools, grain, and agricultural inputs. They already have business dealings with farmers and some often buy grain at harvest time for later resale for consumption or planting when prices are higher. They could diversify into seeds, for instance, by serving as sales agents for public or private companies. In this way they could reduce business risks, acquire experience in handling seed and become a good source of improved seed for farmers who do not have direct access to certified seed from the formal sector. Since traders already operate viable businesses in the community, they are able to generate cash flow for investment in seed production and are able to provide farmers with credit or cash payment for their produce. Traders will become more willing to engage in seed trade if they are provided appropriate incentives including training. van Santen and Heriyanto (1996) describe how local traders have become involved in the supply of soya bean seed in Indonesia. Other examples include the sale of maize and cowpea seed in Ghana (Lyon and Afikorah-Danquah, 1998) and hybrid seed of maize, sorghum, pearl millet and sunflower in India (Chopra, 2000).

Cooperatives and Farmer Organizations

Cooperatives and associations are usual ways of bringing farmers together for mobilization of resources and collective action. They enhance the access of farmers to inputs, credit, extension advice, training and market for their produce. The inclusion of seed in the range of products and services that cooperatives provide could help in making new varieties available to farmers. Some key farmers within the cooperative can serve as lead producers of seed, which will be distributed to members of the cooperative and other farmers. The role of cooperatives in pearl millet in Namibia (Lechner, 1997) and farmer groups for beans in Uganda (David and Kazozi, 1999) illustrate the importance of local organizations in producing and marketing seed of low profitable crops.

Non-Governmental Organizations

Since the 1970s, NGOs have become active in the supply of seed primarily during emergency situations such as drought, floods and civil wars. These organizations have a comparative advantage in providing emergency relief because of the community-oriented nature of their operations. There have been attempts by NGOs to link relief to development so that communities prone to disaster become better prepared to cope with such conditions when they recur. Part of this strategy has been to encourage the development of small seed businesses that can meet the variety and seed needs for specific areas, as in the cases of international NGOs in The Gambia (Cromwell et al., 1993) and ActionAid in Malawi (Musopole, 2000).

Small-scale enterprises of these various types have obvious advantages over large seed organizations in serving the needs of small farmers in scattered locations who demand small quantities of seed of particular varieties for specific conditions. It is difficult for large organizations to serve such needs, since they usually prefer to concentrate on few varieties with wide adaptation and cover large areas of land. In particular, moving small quantities of seed over long distances in remote areas with poor infrastructure will involve high transport costs. The comparative advantages of small enterprises can be explained in terms of unique features that they possess as outlined in the following section.

ESSENTIAL FEATURES OF SMALL-SCALE SEED ENTERPRISES

The success of any enterprise is determined largely by its internal structure and the relationship it has with other relevant institutions and customers. This is particularly true for businesses with small turnover and whose success depends mainly on close and trustworthy relationships with their customers. Small-scale enterprises that produce and sell quality seed in rural farming communities of developing countries belong to this category.

Simple Organizational Structure

Since small enterprises depend on few people, these must have a broad knowledge of all operations, since they are involved in a wide range of activities. The organizational structure should be simple, with few lines of command. The small size and structure will enable direct supervision and intervention, quick decision-making and flexibility without reference to a long chain of command. Kugbei and Turner (2000) suggest a simple organizational structure for a small enterprise managed by three individuals as illustrated in Figure 2, where the enterprise is managed by the owner or manager and two assistants, who are either family relatives or hired persons. One of these supervises field multiplication, processing and storage, while the other takes care of promotion, distribution and sales. These are direct areas of responsibility but the three key staff members work together as a team and understand each other's functions very well.

FIGURE 2. Possible structure of a small-seed enterprise

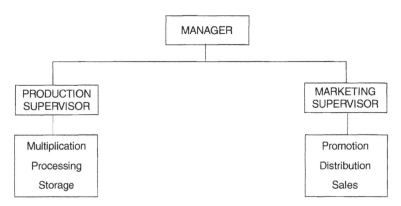

Proximity to the Customers

In many countries, poor infrastructure development particularly internal transportation and communication can lead to a physical isolation of communities and farmers in remote areas. It then becomes difficult to provide timely seed delivery from external sources at prices farmers can afford and are willing to pay. The alternative is to devolve seed production and supply by involving the communities themselves and focusing on their immediate needs. Such production system will better match seed supply with actual demand especially when farmers and extension agents are closely involved, and will facilitate the use of varieties that are recommended for specific locations. The proximity of small seed enterprises to the customers does not only ensure timely availability, but could also play an important role in increasing farmer awareness about the use of improved varieties and seeds within the local community, and hence influence the process of adoption.

Technically Efficient and Cost Effective Operations

Seed costs more because of the additional care and attention given to its production than that of grain of the same crop. Since small farmers generally save their own seed, often as a by-product of grain production, any seed coming from external sources has to be of clearly better quality to justify spending more money on it. The assessment of such quality is easier in a situation where the producer and seed user are in close proximity to one another and may even visit each other's farm during the cropping season. The producers will therefore strive to keep high standards since their operations cannot be hidden from other farmers within the community. Enterprises that serve the seed needs of small farmers in location-specific conditions are usually small, because the markets they cater for are small. Since such enterprises normally deal with less profitable crops, the margins they earn are also small. It means that to generate any meaningful profit, these enterprises have to keep each cost step as low as possible and maintain reputation in the community as reliable and consistent suppliers of good seed, as well as the source of new better-performing varieties. In particular, those enterprises that produce seed of self-pollinating crops cannot expect to offer a sales price that is much higher than the price of grain of the same crop, since the farmer has his own seed as a close substitute. It is a challenge for enterprises to produce seed of good quality at low cost and to sell the final product at a reasonable price. Only small enterprises with low overhead costs have the potential of doing so. Even in such cases, it is difficult for enterprises to survive on seed sales alone. Diversification into other related business areas such as grain trading and input sales is necessary for financial viability. The farmer seed grower is a special case in which success depends to a large extent on efforts to minimize costs and risks.

For self-pollinating crops, the sales price of seed (P_s) can be represented as follows:

$$P_s = P_g + x\% + y\% + z\%$$
$$= 100\% + x\% + y\% + z\%$$

where:

P_g = the price of grain, which is taken as the basic production cost and equivalent to 100%,

$x\%$ = premium paid to the seed grower,

$y\%$ = enterprise's cost margin, comprising the costs of raw seed procurement, processing, storage, transport and all overheads, and

$z\%$ = enterprise's profit margin.

This basic cost structure holds for any enterprise, but the key questions are what $100\% + x\% + y\% + z\%$ equals to for a given enterprise, and what the farmers can afford to pay? The key component is the cost margin ($y\%$), which is often high for a large enterprise but reasonably low for a small enterprise having simple equipment, reduced transport and staff costs.

Close Social Interactions

Small seed enterprises in developing countries are usually household and family-level businesses. They fulfill a social role since they are more labor-intensive than large seed companies, and therefore contribute to job creation and improved livelihood of the poor in the rural areas. To operate effectively in rural farming communities, these enterprises require intimate contacts and a good knowledge of local customs, beliefs, perceptions and norms, since the needs of small farmers must be understood within these contexts. They also serve an extension function by helping promote awareness about new varieties and the use of quality seed.

ENTERPRISE SUPPORT SERVICES

If private enterprises are to emerge as seed suppliers, then appropriate policies are to be formulated and business development services are to be provided as an incentive. Since small-scale enterprises should be capable of serving small, a thorough background analysis is necessary for the identification of appropriate policy, regulatory, technological and organizational interventions that could accelerate the diffusion of new varieties and seeds. Of particular importance are establishing effective links between formal and informal seed sectors, and finding appropriate external means of coordinating the activities of alternative seed supply institutions within the seed industry. Support services for enterprise development can be divided into two phases. First, are those services that are needed as a prerequisite for starting the business, and those that are required to support continued growth after the establishment phase. Start-up support services could take several forms, which are described below.

Policy Reform

Appropriate policy is a fundamental support that entrepreneurs need to establish and develop their businesses. Potential entrepreneurs need to be motivated to enter the seed industry and establish new private enterprises. To do so, any major barriers to entry have to be removed. Such barriers are normally imposed by government policies, some of which need to be revised, or removed and replaced by new ones. The objective is to promote private investment and to create an enabling environment within which businesses can operate. This environment has economic, physical and institutional dimensions, which should all not impede the growth and survival of business.

Enterprise Legislation

Enactment of a favorable seed legislation that recognizes the role of small enterprises is essential as a show of commitment by the government to diversify seed supply in favor of small farmers. Small enterprises need to have a recognised legal status as this will determine the manner in which financial transactions are conducted including tax payments, and access to essential services such as credit and savings. It is important for entrepreneurs to become fully aware that they are setting up enterprises within appropriate and recognised legal frameworks that protect their business interests, particularly investments they make, property rights and enforcement of contractual obligations. Specifying the legal basis of enterprise development can help entrepreneurs in assessing possible risks of financial failure and enable them to find ways of mitigating such risks. This fits well into the overall context of rural business development.

Institutional Support

The development of small seed enterprises should be part of an overall government strategy in rural development. Policy and regulatory interventions alone will not be sufficient to stimulate the development of small seed enterprises. What is need, in addition, is a concerted effort in supporting and promoting enterprises based on specific variety and seed needs of rural communities. Local markets and farmer-to-farmer exchange are essential elements of sustainable seed systems in rural communities. If enterprises are to operate as business entities in their own right, it is important that they are adequately linked to related formal sector institutions such as foundation seed supply, multinational sources of seed and certification service if required. The role of lead institutions in coordinating and supporting local enterprises must be understood clearly, particularly in raising awareness through relevant publications, and organizing events to sensitize policy makers and farmers.

Credit

Access to loan finance, improved technology and information are crucial to success in enterprise development, particularly in the initial stages. The magnitude of start-up capital could be substantial and beyond the immediate reach of potential small-scale entrepreneurs who would be interested in establishing enterprises. For such purposes, favorable loans or credit to acquire inputs, basic equipment and facilities for land preparation, seed cleaning and storage are essential. Loans from formal sources such as commercial banks are usually scarce, particularly for small farmers who may be considered a credit risk. Rural or agricultural development banks and non-governmental organizations usually play a valid role here.

Temporary Subsidies and Tax-Free Holiday

Following initial establishment, small enterprises may need some time to grow and gain financial strength. Since the operational margin is normally small, temporary subsidies and a grace period without tax payment would enable them to accumulate sufficient reserves and be in a better position to survive later when these services are withdrawn.

Quality Control and Regulation

There are limitations in attempting to impose quality control blueprints directly from elsewhere because of differences in local conditions. Although the absence of effective quality control may constrain seed delivery at the local level, it is difficult to implement timely measures in decentralized and scattered production environments. Since resources may also be limited, it is important to collect, document and understand farmers' seed quality problems and traditional management practices in order to design and introduce appropriate crop specific guidelines for quality seed production. Small enterprises need time to gather experience and develop confidence. A quality control system can be introduced for enforcing standards, regulations and codes of practice, as well as for minimizing malpractice. This system should be flexible, impartial and based on simple, realistic and achievable standards. It should provide protection both for the seed producer and for the user. This should include a registration or licensing system for enterprises that meet specified criteria and are accepted as seed producers. A government institution best provides these various functions.

The establishment phase should be followed by a second phase during which the enterprise grows and expand. This stage of development requires certain services that are more lasting in nature and support the existence of the enterprise in the long term. They are described below.

Provision of Appropriate Cleaning and Storage Facilities

Small seed enterprises should produce seed of acceptable quality if they have to attract customers from the community. At the local level, farmers usually clean their own seed using traditional manual methods which are normally suitable only for small quantities. Local enterprises that produce seed for larger markets will require better facilities such as small-scale manually operated cleaners and seed treating equipment. The design of these simple machines should be based on existing experience and indigenous knowledge. Where possible, use should be made of local manufacturing companies that fabricate similar equipment or tools in order to avoid the import of expensive and sophisticated ones. Although the quantity of seed produced is often small, it is necessary to maintain a high quality status and provide protection against storage pests until the time of sales. Simple metal bins have been found to provide safe and effective seed storage at the local level (Osborn and Faye, 1991).

Continued Access to New Varieties and Early-Generation Seed

Seed enterprises must be closely linked to sources of new varieties and early-generation seed so that they are in a position to offer sufficient choice to farmers. The usual sources are agricultural research stations and seed companies or parastatals in the formal sector. Here too, the government is best suited to provide such service until the enterprises become strong enough to maintain their own independent links. In recent years, participatory plant breeding has become an alternative variety development approach for less favorable environments where small-scale farmers predominate (Turner and Bishaw, 2000). Agro-ecological diversity and location-specific preferences of farmers require the release of many varieties with distinct features adaptable to specific conditions. Farmer participatory selection has to be combined with decentralized seed multiplication during the early stages if these varieties are to be adopted and diffused on a wider scale. Therefore, links with small seed enterprises is extremely essential to scale-up seed multiplication if the benefits of participatory plant breeding are to be exploited to the optimum.

Training

Training is an on-going activity and should cover a wide range of subject areas including seed agronomy, post-harvest handling, marketing and business management. The training approach should be participatory and build on indigenous knowledge or locally developed systems as fundamental source of information. Training modules should be designed based on crop-specific issues and local knowledge. They should be practically oriented, easy to understand, and give less emphasis to theory. The beneficiaries should include the staff of seed enterprises, extension staff and farmers.

Increasing Market Potential

The more successful are enterprises in diffusing new varieties, the greater market share they can obtain, which will depend on farmers becoming aware of the attributes of new varieties and seed that the enterprises provide. The farmers need to be convinced that these can bring them more benefit than the old varieties and seed they use currently. This will require efforts in extension and promotion, which could be expensive particularly for enterprises in the initial stages of development. External assistance may be sought from government departments, development agencies or non-governmental organizations working in the community.

Information on National Seed Sector

National seed industries are undergoing significant changes as many of them shift from public to private means of seed supply. Farming communities, therefore, need access to valid information on policy and opportunities that are available for enterprise development. Data collection is costly and it is necessary that the public sector provides the lead in collecting valid information on varieties, seed supply and business development opportunities, which should be analyzed and the results made equally available to all interested parties.

Stimulating Seed Use by Farmers

Small enterprises can only survive if there is enough seed demand in the farming community. Small farmers may wish to try improved seed but encounter constraints in doing so. Some may be short of cash particularly at planting time when they have sold all their produce to repay the previous year's debts, and are left with limited choice including the use of seed they have saved on the farm. Credit facilities to purchase seed of new varieties and accompanying inputs may be absent. Farmers may also be reluctant to adopt new technology because input prices are high while producer prices for grain are low. These will have negative effects on farm incomes and hinder investment in seed production. Governments therefore need to formulate strategies that favor both seed production and use. Important aspects of such a strategy would include the removal of price distortions in the grain and input markets, since these markets are the basis of seed production costs and the final selling price of seed.

CONCLUSIONS

Considering the limited success of the public sector in delivering seed to small farmers in remote areas, and the lack of commercial interest on the part of large seed companies, it seems that small and location-specific enterprises may be the best option for fulfilling this role. However, the strategy in promoting such enterprises should differ from the usual 'top-down' measures characteristic of large seed projects and companies because the conditions are different in terms of crops, resources and potential profitability.

Small-scale enterprises require community-based interventions, since these businesses are meant to serve small farmers in rural areas. Attracting investment in this form of seed supply and creating sufficient interest in the community for the seed the enterprises produce are formidable challenges. They need a strong commitment on the part of governments in introducing favorable policies and providing adequate incentives that can encourage investment.

The enterprises should be allowed to evolve using the community as a firm basis for this process. Therefore, any policy interventions should have lasting effects and point in the right direction of providing adequate support and protection for both the producers and the seed using customers. The aim is to ensure that pioneer enterprises become successful by operating on their own without external support, particularly in the form of operational subsidies although it is important to make the environment favorable and subsidize services if necessary. Once this happens, more investors will enter the seed supply business thus creating a competitive atmosphere that will enhance quality and keep prices at reasonable levels. The production and use of seed will then become sustainable since forces within the farming community will drive both supply and demand. There may be a need at a later stage to organize successful small enterprises into some form of association as a means of protecting their interests and serving as a forum for sharing information as well as exchanging ideas, experiences and lessons.

REFERENCES

Bishaw, Z. and S. Kugbei, 1997. Seed Supply in the WANA Region: Status and Constraints, pp. 71-79. In: Rohrbach, D.D., Z. Bishaw and A.J.G. van Gastel (eds.). *Alternative Strategies for Smallholder Seed Supply*. Proceedings of an International Conference on Options for Strengthening National and Regional Seed Systems in Africa and West Asia, ICRISAT, Andra Pradesh, India.

Chopra, R.K., 2000. Contract Production and Marketing by Prithvi Seeds in Andra Padresh, India, pp. 84-97. In: Kugbei, S., M. Turner and P. Witthaut (eds.) *Finance and Management of Small-Scale Seed Enterprises*. ICARDA, Aleppo, Syria.

Cromwell, C. and S. Wiggins, with S. Wentzel, 1993. *Sowing Beyond the State: NGOs and Seed Supply in Developing Countries*. ODI, London, UK.

David, S. and S. Kazozi, 1999. Designing Sustainable, Commercial and Farmer Seed Production Systems in Africa: Case Studies from Uganda, pp. 128-140. In: Fujsaka, S. and A. Jones (eds.). *Systems and Farmer Participatory Research: Developments in Research on Natural Resource Management*. CIAT, Cali, Colombia.

Kugbei, S., 2000. *Seed Economics: Commercial Considerations for Enterprise Management in Developing Countries*. ICARDA, Aleppo, Syria, viii + 182 p.

Lechner, W.R., 1997. Seed Management and Distribution through Farmers' Cooperatives in Namibia, pp. 135-138. In: Rohrbach, D.D., Z. Bishaw and A.J.G. van Gastel (ed.). *Alternative Strategies for Smallholder Seed Supply*. Proceedings of an International Conference on Options for Strengthening National and Regional Seed Systems in Africa and West Asia, ICRISAT, Andra Pradesh, India.

Lyon, F. and S. Afikorah-Danquah, 1998. *Small-Scale Seed Provision in Ghana: Social Relations, Contracts and Institutions for Micro-Enterprise Development*. Agricultural Research and Extension Network Paper, 84. ODI, London, UK.

Musopole, E., 2000. Small-Scale Seed Production and Marketing in Malawi: The Case of a Smallholder Seed Development Project, pp. 78-83. In: Kugbei, S., M. Turner and P. Witthaut (eds.) (2000). *Finance and Management of Small-Scale Seed Enterprises*. ICARDA, Aleppo, Syria.

Osborn, T. and A. Faye, 1991. *Using Farmer Participatory Research to Improve Seed and Food Grain Production in Senegal*. Development Studies Paper Series. Winrock International Institute for Agricultural Development, Morrilton, USA.

Turner, M.R., 1999. The Prospects for Privatisation of the Seed Sector in Developing Countries, pp. 241-256. In: *Proceedings of the World Seed Conference*, Cambridge, 6-8 September 1999. ISTA, Zurich, Switzerland.

Turner, M.R., S. Kugbei, and Z. Bishaw, 2000. Privatisation of the Seed Sector in the Near East and North Africa. In: *Seed Policy and Programmes in the Near East and North Africa*. FAO Plant Production and Protection Paper 159, pp. 125-136. FAO, Rome, Italy.

Turner, M.R. and Z. Bishaw, 2000. *Linking Participatory Plant Breeding to the Seed Supply System*. Paper presented at international workshop on Scientific Basis of Participatory Plant Breeding and Conservation of Genetic Resources held from 8-14 October 2000 in Oaxtepec, Mexico.

van Santen, C.E. and Heriyanto, 1996. The Sources of Farmers' Soybean Seed in Indonesia, pp. 113-137. In: van Amstel, H., J.W.T. Bottema and C.E. van Santen (eds.). Integrating Seed Systems for Annual Food Crops. Proceedings of a workshop held in Malang, Indonesia, 24-27 October 1995. CGPRT Center, Bogor, Indonesia.

Challenges and Limitations of the Market

Cees van der Meer

SUMMARY. The past decades have shown a shift of paradigm from governments as the main engine of rural development to markets and private initiative. Broad-based growth and better functioning markets are the main hope for the 800 million poor people living in rural areas. Reliance on government services only has proved to be inadequate. Sole reliance on private initiative and markets only is also deficient, given the many market failures, in particular those faced by the poor, and shortcomings in availability of public goods. Therefore, without an effective pro-poor strategy for the private sector in rural development, the efforts in poverty reduction will doubtless be ineffective. *[Article copies available for a fee from The Haworth Document Delivery Service: 1-800-HAWORTH. E-mail address: <getinfo@haworthpressinc.com> Website: <http://www. HaworthPress.com> © 2002 by The Haworth Press, Inc. All rights reserved.]*

KEYWORDS. Seed enterprise development, public-private, seed markets, market failure

The importance of private-sector development was recognized in the World Bank's comprehensive rural development strategy, *Rural Development: From Vision to Action* (VtoA) adopted in early 1997 (World Bank 1997). The strategy puts poverty reduction in rural areas, improvement in well-being of rural people, and the elimination of hunger as the main strategic objectives in the

---

Cees van der Meer is affiliated with World Bank, Washington, DC.

Address correspondence to: Cees van der Meer, The World Bank, 1818 H Street, N.W., Washington, DC 20433.

[Haworth co-indexing entry note]: "Challenges and Limitations of the Market." van der Meer, Cees. Co-published simultaneously in *Journal of New Seeds* (Food Products Press, an imprint of The Haworth Press, Inc.) Vol. 4, No. 1/2, 2002, pp. 65-75; and: *Seed Policy, Legislation and Law: Widening a Narrow Focus* (ed: Niels P. Louwaars) Food Products Press, an imprint of The Haworth Press, Inc., 2002, pp. 65-75. Single or multiple copies of this article are available for a fee from The Haworth Document Delivery Service [1-800-HAWORTH 9:00 a.m. - 5:00 p.m. (EST). E-mail address: getinfo@haworthpressinc.com].

Bank's rural-development efforts. Widely shared growth is seen as an important goal for which private-sector development is one of the strategic principles. "Vision to Action" notes as a lesson learned from past experience that production, input supply, processing and marketing are best carried out by the private sector.

The private sector in broad terms includes all for-profit activities. It consists of micro, small, medium and large enterprises, cooperatives and private-sector organizations. Private-sector players in seed markets range from multinationals and their subsidiaries, to local seed companies, seed multipliers, seed traders, farmers, cooperatives and many kinds of licensed and unlicensed traders. Many of them are small and medium companies, competing in local seed markets. There are also public seed companies and research organizations in some market segments, but their market share is declining with market development and privatization policies.

Most countries developed a dominant government role in plant breeding, variety improvement, seed production and distribution. Initially, when breeding efforts started on a significant scale–for most countries this was only in the 1960s and 70s–there was hardly any or no private seed industry at all, but later, entrance of foreign companies and development of local companies was often strongly restricted by regulation. They also regularly faced competition by public agencies and parastatals. Public seed systems in many countries have proven inefficient and ineffective and in many countries efforts are being made to develop a new vision and strategy for private sector development in the seed sector and a new balance and synergy between government and private sector roles. This often involves painful reforms.

This paper discusses what markets and the private sector can contribute and what their limitations are. In that context it discusses what governments could and should do to let seed markets function well for agricultural growth and poverty alleviation. This contribution draws heavily on a recent World Bank study on good practices in input regulations by Gisselquist and van der Meer (2001), and on a paper by DFID (2000) on markets and the poor.

IMPORTANCE OF SEED MARKETS

Competitive seed markets are very important to agricultural development. Ideally, they provide farmers with new and better varieties at reasonable prices. Open markets provide access to the best available technology and also induce competition between players in the seed market. Farmers are always interested in better performing seed at the lowest possible prices. Plant breeders, seed companies, seed multipliers, wholesale and retail traders are all seeking to satisfy the preferences of farmers, and–behind them–processors and consumers who are buying farm products.

The general opinion among economists and policy makers is that markets driven by private sector interests provide better services to farmers and consumers, and more dynamics to the sector than government-dominated systems of plant breeding, variety development, seed multiplication and distribution. There is also much empirical evidence from OECD and developing countries that such seed market systems can perform efficiently and effectively. The role of the public sector in most OECD countries is mainly to facilitate the performance of markets and the private sector through regulations and supportive research, whereas the role of the government as producer of varieties and seed for sale has declined. Distribution is left entirely to the private sector. In developing countries markets are still less developed and governments play important roles.

Seed markets in most developing countries are underdeveloped and fragile. In many areas and for many crops, formal seed markets hardly play a role. Most seed is still produced by farmers themselves, or locally obtained from other farmers. These local seed systems may function reasonably well for normal circumstances, but their performance is weak in introducing new technology and providing seed in case of emergencies, such as drought or civil disorder (Tripp, 2000). For this reason, government agencies and NGOs have tried to intervene in these systems (Cromwell, 1985), and private companies have some interest to expand their market activities in these areas. Market development is slow, in particular in the case of most food crops. Market penetration has been best in cases of highly performing hybrid varieties, such as maize, and certain vegetables. Farmers want such seed because of its performance and because they cannot produce it themselves, they have to purchase it annually from the companies that owns the parental lines.

But, also in developing countries there are many examples of thriving seed markets, especially for horticultural crops, but increasingly also for industrial and food crops. There are major differences among countries and between crops, depending on policies adopted during the past decades, the regulatory framework, the degree of development of the rural economy, the size of seed markets, and the availability of hybrid varieties. There is clear evidence that countries that opened their markets have seen a rapid development of private sector activities, whereas in other countries private sector activities are still stifled by restrictions (Gisselquist and van der Meer, 2001).

Obviously, market development is a gradual process. The shift from a government driven approach to development to a market and private sector driven approach, does not immediately result in competitive markets. And as we will show below, some problems will never be solved by markets in a satisfactory way without public support. In a rural development strategy there are, therefore, two important questions related to seed markets. First, given the fact that seed markets are underdeveloped and imperfect, what can be done to help farmers get access to better seed, and second, what kind of conditions and interventions are necessary and useful to enhance development of seed markets. Governments have often tried to compensate for market failures by setting-up direct intervention through government agencies, rather than helping the development of competitive markets and a competitive private sector. The way this was done not rarely delayed the emergence of better functioning markets. On the other hand, there are also approaches to market development that may work in the long run, but do not contribute to the solution of problems farmers and seed companies are facing in the short run. Therefore, it is important to clarify what factors and conditions will help to develop competitive markets. In this context special attention has to be given to market imperfections and the role of government.

CONDITIONS FOR SEED MARKET DEVELOPMENT

Basic to the functioning of markets are transaction costs. If the transactions costs in the chain from breeder to farmer are too high, the seed will be too expensive, farmers will not be interested in buying it, local shopkeepers will not put it in their stores, etc. Many factors can cause high transaction costs. Transport costs are high when there are no roads and efficient transport facilities. Small volumes of trade will lead to higher cost of transportation, less competition and higher trade margins. All kinds of government regulations and administrative requirements can increase the cost to companies of introducing new varieties in the market. In many countries requirements for variety registration, and approval for import and selling seed, are too time-consuming, burdensome and costly (Gisselquist and van der Meer, 2001). There are also many kinds of risks that raise the costs and form an obstacle for market development. Farmers may have insufficient knowledge about the performance of new varieties, or be uncertain about the quality of seed offered in the market. Also, seed suppliers may find it hard to get sufficient information on highly diverse needs of farmers, especially those producing mainly for local markets under marginal agro-ecological conditions. Traders may face occasional competition from extension agencies and NGOs that distribute seed below cost price, which leaves them with unsold stocks, or prices too low to cover their costs. Government policies and interventions in other fields may also result in business risk, such as sudden change in policies and regulations in fields of business laws, price controls, taxation, approval procedures, and performance of mandatory public services.

Competitive markets require a sufficient number of competing large and small companies in retail, wholesale and production. Large companies can provide links to foreign markets, technology and capital, and have strong logistic capabilities, but they have high overhead costs. Local shops, small local seed companies and farmers groups are often more efficient in seed multiplication and retailing because they have better knowledge of local conditions and lower overhead cost. Competitive seed markets, therefore, will consist of small and big players that cooperate and compete. The emergence and growth of competitive companies requires skilled employees, managers and entrepreneurs, and capital, and therefore takes time.

The performance of new varieties and the knowledge and experience of farmers are also important factors for demand. If new technology is not superior it will be difficult to introduce. Performance must be clear to farmers. And farmers who never bought better seeds in the formal market may be slower to adopt new varieties than farmers who are in touch with seed markets. A market-oriented agricultural sector will have much closer links to seed markets than a sector that only serves local food requirements. Therefore, the development of seed markets and market-oriented agriculture are inter-linked.

The development of markets and private companies is primarily the result of the activities of private sector players, i.e., farmers, local shops and traders, local seed companies, and big national and multinational companies. However, the government has many roles to play in the development and functioning of competitive seed markets. Private-sector players are faced with many market failures that can be solved or mitigated by public sector interventions.

LIMITATIONS TO MARKETS: MARKET FAILURE

The term 'market failure' identifies the following types of market failure (for a more detailed discussion see DFID, 2000):

- *Public goods*, which the private sector will not supply (or will under-supply) because it cannot appropriate the benefits, such as basic infrastructure, commercial laws, seed laws, law enforcement, and many varieties.
- *Externalities*, which exist when the production or consumption of a good or service has spill-over effects not reflected in the market price, such as in control of contagious seed-born plant diseases, or in marketing better seeds which the farmers can multiply themselves. Selling such seed to farmers may have important spill-over effects to other farmers.
- *Market power and economies of scale*, where barriers to entry create market power, enabling monopoly rents to be earned thus depressing production. Privatization of seed parastatals without foreign competi-

tion, or entry of one international enterprise in a small seed market, may lead to monopoly power.

- *Asymmetric information*, where parties to a transaction have different information about the nature of the exchange. Information failures are particularly widespread in seed markets. Seed companies know more about performance of different varieties and the quality of particular seed shipment, and farmers may find out only when it is too late.
- *Cost of establishing and enforcing agreements* may be so high as to increase risks to the point at which markets do not exist. The cost for individual farmers to get compensation from seed sellers for ill-performing seed may be prohibitive under normal commercial law.

Correcting for market failures provides one widely accepted justification for market intervention. This will often be done by the state, but also the private-sector or cooperative agencies. Seed associations and producer organizations do play important roles in several countries, often involving self-regulation or other collective action, as well as public-private arrangements (Le Buanec, this volume). Actions to correct for market failures are often warranted, provided that state failure or the costs of intervention do not outweigh the original market shortcoming. Means of state intervention in the seed sector include first of all, regulation (including intellectual property rights), funding of research and extension and provision of services and infrastructure.

RESOLVING MARKET FAILURE

Arguments about whether or not markets have failed and, if so, about whether and how governments could intervene to improve the situation, are basic to regulatory design. These arguments are sometimes clear and decisive, but certainly not always. Therefore, good analysis is needed before decisions can be made. The first challenge is to design effective and cost efficient regulations that avoid losses from over-regulation as well as under-regulation. It is useful to consider the following questions:

- Will regulations achieve the objective?
- Will regulations have other (unwanted) effects?
- Is the value of risk or loss averted significantly less than the total cost of regulation (including not only government costs but also impact on business expenses, new entry, technology transfer, and so on)?
- Are there more cost-effective alternatives for achieving the same aim? For example, funding public research may be a more effective form of infant industry promotion than establishing trade barriers.

There is not one blueprint for seed regulation that will give the best results in all cases. On the one hand, regulations should be designed with close attention to common practices in countries that already have competitive markets and industries in place. But an optimal design will also depend on local conditions in each country. Factors to consider include:

- Size of the market.
- General level of development.
- Potential for reliable law enforcement and judicial decisions (transparency of regulatory agencies and decisions, farmer and trader access to judicial processes, etc.).
- Existence and functioning of trade and industry groups.
- Crops and the ecosystem.
- Existing regulations.
- International agreements that limit regulatory options at the national level (Le Buanec & Heffer, this volume).
- Regulations in other countries that may be important partners for seed and fertilizer trade (including especially neighboring and other regional countries).
- The potential interest of private companies.

Market size has a major impact on what are workable regulations for competitive seed industries and trade. Most developing countries are too small to support competitive seed industries focused on the domestic market alone. In the EU, India, or the US, seed and fertilizer markets cross national and state borders to create large unified markets that allow competition, economies of scale, and investment in research. Comparably large markets could evolve in Central Asia and Africa, for example, with regulations and other policies that allow seed industries to operate across regions with minimal time and money lost in moving new varieties and seeds from one country to another.

GOOD PRACTICE IN SEED REGULATION

In their recent good practice guide for the World Bank, Gisselquist and van der Meer (2001) recommend the following principles for regulation.

- For each crop, farmers need a choice of seed from a number of competing companies if markets are to be competitive. Important for competition is that new suppliers can easily enter the market. SME are necessary to ensure that seed industries cover all crops, including crops such as wheat, rice, cassava and local vegetables with low value non-hybrid seeds. The challenge is to accommodate SME seed companies with workable seed regulations (Kugbei and Bishaw, this volume). With reasonable regulations, many seed-producers and traders in what is now known as the "informal sector" can enter the formal sector.

- Governments are recommended to give priority in their efforts to strict truth-in-labeling regulations (possibly in combination with minimum standards for seed quality). In principle, regulations should allow companies to sell seeds of new varieties without having to obtain approval from ministries of agriculture. There could be some exceptions. Countries intending to join the EU should accommodate for the rules governing the EU Common Catalogues. For a few major food and export crops, companies could be asked to register varieties and to show the results of performance tests before seed can be freely sold, but costs and benefits of such requirements should be carefully considered.

- Governments are recommended to allow seed companies and traders to operate without licensing or with near-automatic licensing from the Ministry of Agriculture. They should allow retail seed sales (at least to some maximum annual value) without licenses. The government can leave the selection of seed farmers completely to the private sector.

- Seed export controls can only be justified in rare cases. To protect agricultural production, governments should control seed imports to block introduction of seed-borne pests and diseases that are not found in the country and that are potentially damaging following IPPC and WTO/SPS. To protect indigenous biodiversity, governments should regulate the introduction of new plant species and agricultural biotechnology to avoid introductions to become invasive. Governments are recommended not to set non-tariff barriers based on quality, quantity, or prices.

- To facilitate market development, seed laws and regulations should mandate government agencies to provide a number of services to seed companies. These services include: (a) providing for seed certification and other seed quality certificates; (b) phytosanitary certificates for exported seed; (c) intellectual property rights for plant varieties (PVP) and biotechnology applications. These services should be available when seed companies ask and pay; they should not be compulsory.

- Governments can manage policy and supervision of regulatory tasks through an office in the Ministry of Agriculture along with three regulatory agencies: a seed testing and certification agency, an agency to administer phytosanitary controls, and an agency to administer plant variety protection. These can all be relatively small, and can sub-contract many activities such as seed tests to others, including companies, scientific institutions, etc. If the Ministry of Agriculture continues to list and control allowed varieties, as with arable and fodder seeds in the EU, the seed office in the Ministry may be responsible for maintaining variety lists. The office may also maintain lists of recommended varieties. Obviously, the rules and their implementation should be executed in a (cost) effective and above all transparent way; some of these tasks and agencies could be shared in regional cooperation.

PUBLIC AND PRIVATE ROLES

It is clear that the government should take care of such public goods as seed laws, law enforcement, fundamental research, protection of public health and the ecosystem, education, basic infrastructure and various other kinds of market failure, and leave private-good activities such as production, storage, transport and trade in seed to the private sector. This is simple and straight-forward, and forms the basis for privatization. But, as noted above, relying only on government and on markets is not enough for two reasons. First, there are many examples where the production of private goods can be very dependent on the availability of public goods. The private sector cannot effectively supply seed in areas without transport. Without protection of intellectual property the marketing of certain seed products may not be commercially feasible.

Second, many goods have shades of public and private goods characteristics. They are hybrid or mixed goods. For example, developing new seed markets may be too costly or too slow if there is no effective cooperative effort by agricultural extension or NGOs. Selling seed in marginal areas, or to poor farmers, may not be sufficiently profitable to the private sector. Developing new varieties for small and underdeveloped markets may be too costly in the absence of support from public breeding. Small seed companies may be dependent on the availability of public services in the field of seed technology, human resource development, etc. If the investment in such mixed goods is left to the private sector there will be under-investment. If it is left to government services the impact of the work may be limited, not market oriented and technically insufficient. There is no standard solution. In some cases, forms of public-private cooperation can be effective and efficient to overcome such market failures. Examples are co-funding schemes for technology and for market development.

The public and private sectors often have complementary expertise and overlapping interest in good regulations and effective implementation. This offers scope for public-private cooperation. Governments that are ready to work with the private sector will have an open door to farmers' organizations as well as seed associations to talk about the full range of common concerns, including research, seed regulation, seed export promotion, and so on. Since government agencies often have insufficient human and financial resources to design and implement good regulations, joint policy advisory committees for designing and implementing regulations and policies, consisting of experts from the public and private sectors, can be very important. In other cases, NGOs may be able to help to organize small-scale producers so that they can work to overcome market failures or high transaction costs.

One of the basic institutions for promoting and maintaining a competitive seed industry are independent private seed associations. Seed associations provide information and services to members and also act as a spokesperson for seed companies in dealings with regulators, legislators, and other government bodies. Such associations have to equally represent the interests of all their members; large and small enterprises, international and niche-market players. In cases where there is an especially well-organized private sector, a Seed Board may be in a position to implement public and collective functions very well and the government needs only to provide oversight authority (van der Meer, forthcoming). The government should create a framework for the Board and make sure that all parties, including producers organizations, are involved, and that the Board does not reduce market access for newcomers.

In many countries public agencies have not only played important roles in breeding, seed production and distribution, but also in policy making and implementation. Experts from these agencies often dominate in committees that decide on approvals and controls. When markets are opened for the private sector there is often a tendency for government agencies to use their authority, and experts to control market entry of seeds and private companies. It is important for the government at that stage to separate policy making from policy execution. In addition, the government should explicitly and transparently guide its institutions, to avoid unfair competition between public agencies and private companies. Public agencies should leave the production of private goods to the private sector, work on full-costing base and not use public funds to compete for market share with private companies.

In many countries the regulatory system is working poorly because of insufficient funding. Since the regulatory functions have public goods aspects, some basic funding from the regular budget will be necessary, but since farmers and businesses are profiting from the services of a good regulatory system, work on full-costing can be funded through registration fees, certification fees and inspection and/or tonnage fees.

POLITICS AND REFORM

Although the private sector has become an important source for production and distribution of agricultural inputs in many countries, there are still many developing and transition countries that have not yet fully utilized these opportunities. To tap these opportunities requires public sector reform, deregulation and liberalization. However, politicians may face political constraints that obstruct adoption of regulations that serve the national interest. These constraints may come from various directions, such as economic conflicts of interest, national security concerns, popular beliefs, and the way the political process works. Reform processes have winners and losers. The winners are the farmers, consumers and new entrants in the market. The losers could be protected industries and parastatals that face new competition and loss of privileges. Government services face reduction of authority and funds, so some staff and regulators may consider deregulation a threat to their jobs. Commonly, in all countries losers oppose reforms. Therefore reforms require vision and leadership and sometimes some compensation or a transition period for losers to adjust to the new situation. In planning for policy and regulatory reforms, political factors and administrative resistance should be dealt with effectively, otherwise reforms may not work.

Finally, discussions about the design of input regulations are too important to be left to regulators and government scientists. If the interests of farmers and consumers are to be given proper weight, it is important to involve politicians and senior government officials next to the various stakeholders themselves. The importance of the contribution of political leadership to assist in the introduction of foreign agricultural technology has been demonstrated many times over in the last 40 years.

REFERENCES

Cromwell, E. and S. Wiggins. 1993. *Sowing Beyond the State, NGOS and Seed Supply in Developing Countries*. Overseas Development Institute, London.

DFID, 2000. Making Markets Work Better for the Poor. A Framework Paper, DFID, London.

Gisselquist, D. and C. van der Meer. 2001. *Regulations for Seed and Fertilizer Markets, A Good Practices Guide for Policy Makers, World Bank*, Washington DC.

Tripp, R., 2000. *Strategies for Seed System Development in Sub-Saharan Africa: A Study of Kenya, Malawi, Zambia, and Zimbabwe*, ICRISAT and ODI, Bulawayo and London.

van der Meer, C., forthcoming. Public-Private Cooperation, Examples from Agricultural Research in the Netherlands, In: Byerlee, D., and R. Echeverria, *Agricultural Research Policy in an Era of Privatization: Experiences from the Developing World*, Wallingford, UK, CABI.

World Bank 1997. *Rural Development: From Vision to Action (VtoA)*, World Bank, Washington, DC.

The Role of International Seed Associations in International Policy Development

Bernard Le Buanec
Patrick Heffer

SUMMARY. The regulatory framework surrounding the seed industry is becoming more and more complex and, from the early 1990s, international negotiations are shifting from purely technical issues to more political ones. This evolution is particularly linked to the development of biotechnology, and discussions on biological diversity and intellectual property protection. In that context, international seed associations have a key role to play to ensure that the industry point of view is well understood and taken into account.

In this article, we review the different global organizations and agreements having an impact on activities of the seed industry, including "new comers" such as the Convention on Biological Diversity and the *Codex alimentarius*. Then, we analyze the strategic role of international seed associations in developing consensus among their members, adopting common position papers, and defending these positions in international fora. We also discuss the work of international associations using concrete examples such as the on-going international initiatives on seed health testing, on adventitious presence of transgenic material in non-GM crops, and on the assessment of essential derivation. *[Article copies available for a fee from The Haworth Document Delivery Service: 1-800-HAWORTH. E-mail address: <getinfo@haworthpressinc.com> Website: <http://www. HaworthPress.com> © 2002 by The Haworth Press, Inc. All rights reserved.]*

Bernard Le Buanec and Patrick Heffer are affiliated with FIS/ASSINSEL, Nyons, Switzerland.

Address correspondence to: Bernard Le Buanec or Patrick Heffer, FIS/ASSINSEL, Chemin du Reposoir 7, 1260 Nyon, Switzerland.

[Haworth co-indexing entry note]: "The Role of International Seed Associations in International Policy Development." Le Buanec, Bernard, and Patrick Heffer. Co-published simultaneously in *Journal of New Seeds* (Food Products Press, an imprint of The Haworth Press, Inc.) Vol. 4, No. 1/2, 2002, pp. 77-87; and: *Seed Policy, Legislation and Law: Widening a Narrow Focus* (ed: Niels P. Louwaars) Food Products Press, an imprint of The Haworth Press, Inc., 2002, pp. 77-87. Single or multiple copies of this article are available for a fee from The Haworth Document Delivery Service [1-800-HAWORTH 9:00 a.m. - 5:00 p.m. (EST). E-mail address: getinfo@haworthpressinc.com].

KEYWORDS. International agreements, seed association

INTRODUCTION

Internationally harmonized regulations, procedures and standards are very conducive to support international trade of seed. Several intergovernmental organizations and international non-governmental and industry organizations are charged by their members to develop such rules, procedures and standards.

Traditionally, the seed industry has been actively involved in intergovernmental fora such as the Organization for Economic Cooperation and Development (OECD), the International Union for the Protection of New Varieties of Plants (UPOV) and the Food and Agriculture Organization of the United Nations (FAO), as well as in international NGOs such as the International Seed Testing Association (ISTA). These organizations are conventional partners. More recently, new comers such as the Codex alimentarius and the Convention on Biological Diversity (CBD) are entering the scene. They are little familiar with the seed industry, but will certainly have an important impact on its activities.

The current context is also characterized by a more and more political environment. Negotiations are shifting from purely technical issues to political ones, in particular due to the development of biotechnology and discussions on biodiversity.

In this context, global seed associations have a key role to play to ensure that the seed industry point of view is well understood and taken into account.

This paper reviews the role of international seed associations in international policy development. For a reason of simplicity, it focuses on the global level. However, the situation at regional level is relatively similar. Most regional intergovernmental bodies have a counterpart regional seed association (Box 1).

THE REGULATORY FRAMEWORK SURROUNDING THE SEED SECTOR

Seed is one of the most regulated products in the world. It is governed by a set of laws and regulations dealing with variety registration, seed certification, plant variety protection, phytosanitary certification and, more recently, related to biosafety and food safety. Some of these regulatory instruments are harmonized at international level. This is the case for certification of seed moving in international seed trade, plant variety protection, phytosanitary certification and, recently, biosafety. These instruments depend from different intergovernmental organizations.

BOX 1. International seed associations

In 2001, there are two global and six regional seed associations in place.

- The International Seed Trade Federation (FIS) and the International Association of Plant Breeders (ASSINSEL) are the two global associations representing, respectively, the seed trade and the plant breeders' community. They act as seed industry representatives in global intergovernmental fora such as OECD, UPOV, IPPC, CGRFA, Codex alimentarius and CBD, and in international NGOs such as ISTA. FIS and ASSINSEL were established, respectively, in 1924 and 1938. In 2000, they had members in 66 developed and developing countries. FIS and ASSINSEL share a common secretariat. They will merge in 2002 and form the International Seed Federation (ISF).

- There are regional seed associations in the following regions:
  - European Union: European Seed Association (ESA);
  - Asia and Pacific: Asian and Pacific Seed Association (APSA);
  - Latin America: Federación Latinoamericana de Asociaciones de Semillistas (FELAS);
  - Africa: African Seed Trade Association (AFSTA);
  - West Asia-North Africa: WANA Seed Network;
  - Central & Eastern Europe: Eastern European Seed Network (EESNET)–to be established in 2001.

These associations act as seed industry representatives in regional intergovernmental bodies.

Organization for Economic Cooperation and Development (OECD)

OECD has established worldwide recognized schemes for the Varietal Certification of Seed Moving in International Trade (known as OECD Seed Schemes). There are seven schemes: on herbage crops, oil crops, cereals, maize and sorghum, beet, vegetables, and subterraneum clover. In 2000, 48 countries participated in at least one of the schemes. These countries represent the most active states involved in international seed trade.

FIS was active in the establishment of the first scheme, on herbage crops, in 1957. Amendments to the schemes are under the responsibility of the annual meeting of representatives of the national designated authorities. FIS is invited to attend the annual meeting in an observer capacity, as seed industry representative.

The annual meeting represents an appropriate place to discuss issues that relate to seed certification in international trade. At present, in addition to technical amendments, discussions mainly focus on the issue of accreditation of seed company personnel to fulfill official duties, and they have to deal with the issue of adventitious presence of transgenic material in non-genetically modified (GM) seed.

International Union for the Protection of New Varieties of Plants (UPOV)

UPOV was founded in 1961. Its purpose is to acknowledge the achievements of breeders of new plant varieties by granting them an exclusive property right on the basis of a set of uniform and clearly defined principles. Protection is given to plant breeders both as an incentive to the development of agriculture, horticulture and forestry, and to safeguard their interests, as improved varieties are a necessary and cost-effective input for achieving food security in a sustainable manner. In 2000, UPOV counted 46 member states.

UPOV has established several bodies: the Council, the Administrative and Legal Committee and Technical Working Parties, which meet regularly. FIS and ASSINSEL have an observer status in all these bodies.

It is interesting to note that UPOV is the conclusion of two international diplomatic conferences, the first having been convened in 1956, further to a wish expressed by ASSINSEL during its congress held in Semmering, Austria in 1956.

Food and Agriculture Organization of the United Nations (FAO)

FAO provides the secretariat for three international agreements that are relevant to the seed industry:

- the International Plant Protection Convention (IPPC),
- the Codex alimentarius, and
- the International Treaty on Genetic Resources for Food and Agriculture (IT/GRFA).

FIS/ASSINSEL has been granted the special consultative status in FAO and all fora linked to it.

International Plant Protection Convention (IPPC)

The IPPC is the international convention that provides the framework for the development and application of harmonized phytosanitary measures and the elaboration of international standards to that effect. It is aimed at promoting international cooperation in controlling pests of plants and plant products and preventing their international spread, and promoting technically justified and transparent phytosanitary measures to avoid unjustified restrictions to international trade.

The IPPC was revised in 1997, but the revised text is not yet into force 3 years later. The revised convention provides for a commission on phytosanitary measures, which is in charge of promoting the full implementation of the convention objectives. The interim commission is composed of all IPPC member states plus some observers such as regional plant protection organizations and trade associations, i.e., FIS.

The IPPC and its related standards are very relevant to international seed trade since phytosanitary requirements, sometimes unjustified, constitute today the main barriers to international seed trade.

Codex Alimentarius

Codex alimentarius is the joint FAO/WHO forum in charge of setting international standards for food products. Its objective is to ensure food safety, while facilitating international trade in food products taking into account consumers' and trade concerns.

It is only recently, with the development of transgenic varieties, that the Codex alimentarius became a relevant forum to the seed industry. Since 1997, FIS and ASSINSEL have attended codex meetings on labeling and safety analysis of food obtained through biotechnology. It is worthwhile noting that, in the frame of negotiations within the Codex Task Force on biotechnology, ASSINSEL has been requested by Codex delegates to table for 2001 a discussion paper on possible application of the "familiarity" concept to the food safety analysis (a concept used for biosafety, but new in the field of food safety).

International Treaty (IT) on Plant Genetic Resources for Food and Agriculture

The IT is an agreement promoting a multilateral system for plant genetic resources for food and agriculture. It is currently under revision in harmony with the Convention on Biological Diversity (CBD) in the frame of the FAO Commission on Genetic Resources for Food and Agriculture (CGRFA). Negotiations, initiated in 1993, are rather difficult. Conflicting issues are: access to genetic resources, benefit sharing and farmers rights.

In order to facilitate the negotiation process, ASSINSEL adopted in 1998 a position on access to genetic resources and sharing of benefits arising from their use. This position was welcomed by most delegations, including developing countries and "green" NGOs. This key contribution was given full recognition in 2000 when Norway proposed an article on benefit sharing based on the ASSINSEL proposal.

Convention on Biological Diversity (CBD)

The CBD is one of the conventions resulting from the Earth Summit held in Rio in 1992. It is aimed at promoting the conservation and sustainable use of genetic resources and the equitable sharing of benefits arising from their use.

The first priority of the convention was on biosafety. This resulted in the adoption of the Cartagena Biosafety Protocol in January 2000, which is the first CBD binding protocol. This protocol is aimed at regulating transboundary movements of Living Modified Organisms (LMOs) that may have adverse effects on the conservation and sustainable use of biological diversity. It is not likely to enter into force before 2002. The Cartagena Protocol will have an important impact on the seed industry, since transgenic seed is the main target.

FIS and ASSINSEL have an observer status at CBD level. For negotiations on the Biosafety Protocol, FIS/ASSINSEL established a task force to review the different options and their consequences, and work on a draft protocol text acceptable to the seed industry. This text was widely circulated to delegates participating in negotiations, and was endorsed by some other industry associations. FIS/ASSINSEL participation in CBD meetings is generally coordinated with other industry and trade associations within an informal coalition: the Global Industry Coalition (GIC).

World Trade Organization (WTO)

The three following WTO agreements are relevant to the seed industry: the Agreement on Trade-Related Aspects of Intellectual Property Rights (TRIPs), the Agreement on the Application of Sanitary and Phytosanitary Measures (SPS) and the Agreement on Technical Barriers to Trade (TBT). However, FIS and ASSINSEL, as any NGO, has no observer status in the organization. Therefore, direct relations with WTO are limited. Any WTO-related position has to be defended by the members who have to contact their respective trade ministries.

In 1999, ASSINSEL contributed a publication on the review of Article 27.3(b) of the TRIPs Agreement, dealing with the protection of plant varieties.

International Seed Testing Association (ISTA)

ISTA is a non-governmental organizations of seed testing laboratories. It was established in 1924, similarly to FIS. Its objective is the harmonization of seed testing methods globally. ISTA is an harmonizing body, not a regulatory one.

FIS is actively involved in ISTA discussions. Recently, FIS obtained from ISTA the possibility for company laboratories to be accredited for testing seed and issuing orange international seed testing lot reports (known as "orange certificates").

In the field of seed health testing, FIS and ISTA having not the same priorities, FIS decided to launch the International Seed Health Initiatives (ISHIs), which are industry initiatives with an active participation of ISTA member laboratories (see Box 2).

BOX 2. International Seed Health Initiative (ISHIs)

The seed industry has a responsibility to secure the delivery to farmers and growers of healthy seed, free from seed-borne disease and to respect international phytosanitary regulations. This leads to the need of a systematic approach. However, we do not have at this time a proper response to that important challenge. Different test methods are currently performed all over the world and the results are not generally accepted. In order to improve the situation, seed companies took initiatives to cooperate and exchange knowledge on seed-borne pathogens and their control. These initiatives are trying to assure healthy seed through cooperative evaluation of the methods to assure reliability, reproducibility and suitability of seed testing. They cooperate with ISTA member laboratories and official national and international regulation and accreditation authorities.

Three International Seed Health Initiatives (ISHIs) have been established:
- the International Seed Health Initiative on Vegetable Crops (ISHI-Veg) in 1994,
- the International Seed Health Initiative on Herbage Crops (ISHI-H) in 1997, and
- the International Seed Health Initiative on Field Crops (ISHI-F) in 1998.

Their objective is to secure the delivery to farmers and growers of healthy seed, free from seed-borne pathogens and to facilitate international movement of seeds, whilst preventing unjustified non-tariff barriers.

The main ISHIs output is a reference manual of seed health testing methods for vegetable seeds (see http://www.worldseed.org).

FIS is working on a possible international standard on seed health testing, to be possibly endorsed by IPPC, so that ISHIs and ISTA validated seed health testing methods be recognized as tools for the phytosanitary certification of seeds.

ISHIs are also involved in Pest Risk Analyses (PRAs) in order to assess whether quarantine requirements for some pests are justified. Recently ISHI-F carried out a PRA on *Erwinia stewartii* in maize seed showing that, in view of recent scientific developments, phytosanitary requirements for *Erwinia* should be amended. Other PRAs on seed-borne diseases are in the pipeline.

THE ROLE OF INTERNATIONAL SEED ASSOCIATIONS

In this context, where regulations are more and more harmonized at global level, international seed associations have a great role to play so that seed related regulations be tailored taking into account expectations and constraints of the seed industry. Their active involvement will ensure that new regulations, or amendments to existing regulations solve more problems than they create. In that respect, seed associations have to develop consensus among their members, adopt common positions, and present and defend these positions.

Develop Consensus Among Members

When a new issue arises at international level, it is tabled in appropriate FIS/ASSINSEL meetings, either by members or by the secretariat. Further to the presentation, it may be decided:

- to take immediate action, if the issue is unanimously considered of major importance;
- not to have any follow-up, if the issue is not considered relevant, or if it falls outside the mandate of the association (from a topic or geographical point of view);
- to ask for a discussion paper, if further information is needed before decision;
- to launch ad hoc initiatives/experiments to get the necessary expertise.

Experience shows that decisions are generally taken unanimously within FIS and ASSINSEL since the members are confronted to the same practical challenges and share the same opinion, be they from the public or private sector, large or small companies, or from a developed or developing country. Vote is required in a few cases only. In these cases, during the past decades, decisions have been taken by strong majority, and always with more than 90% of the voting rights.

International seed associations may also establish, on a case-by-case basis, initiatives/experiments to gain experience and solve specific problems. In that respect, three seed industry initiatives/experiments may be quoted:

- The International Seed Health Initiatives (ISHIs): their objective is to develop and harmonize seed health testing methods, and promote their recognition as tools for the phytosanitary certification of seeds. They also carry out Pest Risk Analyses (PRAs) to propose amendments to some quarantine requirements (see Box 2).
- The International Seed Network Initiative (ISNI) on the Transboundary Movement of Seed and Biotechnology: it is aimed at finding appropriate solutions to the issue of adventitious presence of transgenic material in non-GM seeds (see Box 3).
- Experiments on essential derivation: their purpose is to develop tools and define thresholds to assess essential derivation (see Box 4).

Adopt Common Position Papers

When an issue is considered of major relevance, the secretariat works together with the appropriate committee on a draft position paper or statement. This document is then circulated to the members, for comments. If all contri-

---

BOX 3. International Seed Network Initiative (ISNI) on the Transboundary Movement of Seed and Biotechnology

Depending on the crops, the seed market is national, regional or global. Interdependence of countries for seed supply is of particular importance for some crops such as maize, rapeseed, cotton, soybean, herbage crops and vegetables. Up to now, following the OECD Seed Schemes, the international movement of seed was quite easy. The development of transgenic crops in some countries has recently rendered the situation much more difficult, with a risk of disruption of the international seed trade for the following two main reasons:

- The authorization for commercial release of transgenic crops is done on a country-by-country basis, leading to very different situations in different countries;

- Due to realities of plant reproduction (cross-pollination) and seed production, low levels of adventitious presence of transgenic material in non-GM seed can occur.

That adventitious presence creates difficulties and several problems have been encountered in Europe in 2000. The only country having officially adopted standards is Switzerland. As soon as May 1999, FIS drew the attention of governments to that issue and proposed, in August 1999, an experiment to develop standards and quality assurance systems to face this new challenge. The project was presented to OECD and ISTA, which agreed to join the initiative, which has been subsequently transformed into the "International Seed Network Initiative." Whilst not yet unanimously accepted by all members of the OECD Seed Schemes, the experiment should start in 2001 in maize, cotton, soybean and rapeseed.

The experiment should define an appropriate sampling and detection protocol for a given adventitious presence threshold. In the meantime, the experiment should elaborate appropriate quality management procedures that would minimize the risk of adventitious presence.

---

butions are consistent, an amended version accommodating all contributions is submitted for adoption to the general assembly. If contributions show discrepancies, the amended version is subject to a second round of comments before submission to the general assembly and, if needed, ad hoc meetings are convened.

In 1998-2000, FIS and ASSINSEL adopted positions or statements on several topics. Here are some examples:

- protection of parental lines (2000);
- adventitious presence of GMOs in non-GM seeds (1999 and 2000);
- DUS testing (1999 and 2000);
- development of new plant varieties and protection of intellectual property (1999);
- distinctness of "converted" varieties (1999);

- variety denomination (1999);
- essential derivation (1998);
- access to plant genetic resources and benefit sharing (1998).

Moreover, for urgent matters, some positions may be adopted by the Executive Committees, making it possible to react rapidly to new issues that require urgent contribution from the industry. This provides the required flexibility for an efficient lobbying body. This procedure proves to be very useful on some issues debated several times a year, such as the accreditation issue and the issue of adventitious presence of GMOs.

Use of fast-track adoption procedures will probably be more and more needed, since it will become more and more difficult to wait for an annual general assembly to adopt position papers, if the seed industry wants to have timely contributions to international negotiations.

Present and Defend Positions in International Fora

When a position is adopted, it is the role of the secretariat and the members to present and defend it. In that respect, seed associations at national, regional

---

BOX 4. Experiments on essential derivation

The 1991 Act of the UPOV Convention has introduced the new concept of essential derivation in its article 14(5). That concept, aimed at preventing plagiarism and misappropriation of protected varieties, is welcome by the plant breeding community. However, as any new concept, it is not easy to implement, juridical approaches and technical tools needing to be developed.

As regards juridical approaches, ASSINSEL has adopted several position papers promoting codes of conduct and arbitration rather than law suits.

As far as technical tools are concerned, ASSINSEL has implemented during the last decade, and is still implementing technical experiments aimed at developing molecular tools and defining thresholds to assess essential derivation. These experiments are on:
- maize: experiments conducted in France, Germany and the USA from 1993 till now; results of these studies are being consolidated within ASSINSEL;
- tomato: 1994-96;
- ryegrass: 1998-99;
- lettuce: 2000-01;
- rapeseed: 2001 onward.

Results of these studies are circulated to UPOV Technical Working Parties and are also published more widely. They will serve as a basis for dispute settlement in essential derivation.

and global level have all a very important and complementary role to play. For an efficient representation of the seed industry, the position has to be presented to regulatory people at the three levels:

- national associations should present it to their appropriate ministries (e.g., agriculture, trade, environment);
- regional associations should present it to their regional or sub-regional bodies (e.g., EC for the European Union, OAU for Africa);
- global associations should present it to relevant global fora (e.g., OECD, UPOV, FAO, CBD).

Without this close relationship between national, regional and global associations, and companies, seed industry positions would not be efficiently defended. This is particularly true in fora such as the CBD, which are not very familiar with agriculture in general, and the seed industry in particular, and are under the pressure of other very active movements.

CONCLUSION

At present, more and more seed related regulations are, or are being harmonized at international level. This requires great attention of the industry, since these regulations will impact their daily activity. Therefore, the industry has to be actively involved in relevant negotiations. The most efficient way is through international associations that will help find consensus and adopt common positions. Then, it is up to their members and their secretariat to present and defend these common positions in appropriate fora.

At global level the seed industry is represented by FIS and ASSINSEL. These two associations are well known and recognized in intergovernmental fora dealing with seed related issues. Their position is constructive and helps adoption of "realistic" regulations and policies taking into account concerns and constraints of the global seed industry. This will also influence national and regional seed associations when defining their own positions.

Active communication with the members, and a flexible decision making process make it possible to international seed associations to react quickly to urgent topics and to adopt common positions in a timely manner. Nevertheless, active involvement of the national, regional and global levels is necessary if the seed industry point of view has to be effectively presented and defended at global level.

Policy Response
to Technological Developments:
The Case of GURTs

Niels P. Louwaars
Bert Visser
Derek Eaton
Jules Beekwilder
Ingrid van der Meer

SUMMARY. Technological developments may require a policy response when the potential effects of such technology contribute to unwanted or unpredictable changes. The introduction of genetic modification triggered policy makers to design a framework for risk assessment and release procedures that may be linked to conventional variety release systems (Traynor & Komen, this volume). Often, technological change reaches the policy level only when problems appear after introduction. In some cases, however, discussions can start even before the technology is ready for the market. A good example of the latter is the Genetic Use Restriction Technology (GURT), which triggered a very intense debate because of its possible use in the production of 'sterile seeds.' This application

Niels P. Louwaars, Jules Beekwilder, Ingrid van der Meer are affiliated with Plant Research International, Wageningen, The Netherlands.

Bert Visser is affiliated with the Centre for Genetic Resource, The Netherlands (CGN), Wageningen, The Netherlands.

Derek Eaton is affiliated with Agricultural Economics Institute (LEI), The Hague, The Netherlands.

Address correspondence to: Niels P. Louwaars, Plant Research International, P.O. Box 16, 6700 AA Wageningen, The Netherlands.

[Haworth co-indexing entry note]: "Policy Response to Technological Developments: The Case of GURTs." Louwaars, Niels P. et al. Co-published simultaneously in *Journal of New Seeds* (Food Products Press, an imprint of The Haworth Press, Inc.) Vol. 4, No. 1/2, 2002, pp. 89-102; and: *Seed Policy, Legislation and Law: Widening a Narrow Focus* (ed: Niels P. Louwaars) Food Products Press, an imprint of The Haworth Press, Inc., 2002, pp. 89-102. Single or multiple copies of this article are available for a fee from The Haworth Document Delivery Service [1-800-HAWORTH 9:00 a.m. - 5:00 p.m. (EST). E-mail address: getinfo@haworthpressinc.com].

was dubbed "terminator technology" in the popular press. GURT is thus an interesting case to analyse the link between technology and policy development. This paper heavily draws upon a study that was prepared by FAO (Visser et al., 2001).

This case illustrates that a wide range of concerns and options are linked with one technological development, and that arguments arise from different policy fields. Analysis thus needs a thorough understanding of the individual opportunities and concerns as well as the linked arguments.

GURT has received an extremely bad name in the international public debate. Very few, however, have seriously thought about possible policy responses and the tools that are available to the policy makers to implement their decisions. This paper intends to clarify both the complexity of such technological developments and it gives some suggestions about dealing with different concerns in the GURT's case. *[Article copies available for a fee from The Haworth Document Delivery Service: 1-800-HAWORTH. E-mail address: <getinfo@haworthpressinc.com> Website: <http://www. HaworthPress.com> © 2002 by The Haworth Press, Inc. All rights reserved.]*

KEYWORDS. Terminator technology, genetic use restriction technology, biotechnology, biosafety, agro-biodiversity

TECHNICAL: WHAT ARE 'GURTS'?

Several technical methods that provide genetic switch mechanisms have been described in recent patent applications. Such mechanisms, which aim to restrict the use of genetic material are named 'Genetic Use Restriction Technologies' (GURTs). The genetic switch can be used to restrict further multiplication by turning the seed sterile (Variety-GURT or V-GURT, or in the popular press: 'terminator technology'), or to control the expression of certain traits (T-GURT).

At least three general V-GURT strategies can be distinguished. The first strategy makes use of induced activation of a disrupter gene that can inhibit embryo formation (Delta & PineLand/USDA concept). This gene is held dormant by a genetic blockade throughout the seed multiplication process. When the seed is treated before sale to end users, a cascade of events leads to expression of the disrupter in the second generation seed: farm-saved seed will not germinate. The technology is not yet operational: proof is lacking as yet for efficient control of the recombinase. Also effective application of the inducer chemicals to the seed in order to avoid 'escapes' has not been reached yet.

Technically, these systems using a recombinase activity have a lot in common with methodologies to arrive at marker-free transformed plants, which is increasingly required by biosafety regulations.

In the second strategy, the breeder applies a chemical throughout multiplication, but stops to do so before selling the seed (Zeneca concept). In this concept a disrupter gene is expressed in the seed by default, resulting in sterile seed. This system works in the laboratory, but needs further work in order to be effective in the field (Kuvshinov et al., 2001). The third strategy focuses on vegetatively reproducing crops like root and tuber crops and ornamentals. In this concept a gene that blocks growth is expressed by default. This causes the cutting not to form adventitious roots, thus avoiding further multiplication. The ability to form roots can be restored by induction of a second gene.

In T-GURT concepts only a genetic trait is switched on or off at will. This can be realised by different strategies: inducible promoters that regulate the expression of a gene, induced gene silencing, or by excision of the transgene using a recombinase (Zuo & Cha, 2000). Potentially, such technologies may be used to switch on genes that increase characteristics like drought tolerance only when drought occurs.

MOTIVES FOR THE DEVELOPMENT OF 'GURT'

Industrial Interest in Relation to Seed Markets

Breeding itself does not generate income-breeders obtain their return on investment through the sale of seed. In most situations, farmers are able to reproduce their seed and need to purchase a small quantity of seed of a new variety in order to benefit from the breeding activities for several years. Breeding companies wish to have a sufficient level of control over plant varieties in order to safeguard their investments in breeding. Intellectual property rights such as Plant Variety Protection or patents are able to increase the return on investment to varying extends, and often at high costs. Biological protection systems like hybrid or V-GURTs offer a better insurance against 'free-riding.' V-GURTs force farmers to purchase seed every season and T-GURTs can generate benefits through control over the inducer (the chemical that has to be sprayed on the crop in order to induce the expression of the T-GURT protected trait).

The technology has the potential to turn less profitable seed crops like self-fertilising cereals, cotton and legumes into commercially interesting products for the seed industry, especially in countries with an ineffective or very expensive intellectual property rights system.

Furthermore, when V-GURT is combined with apomixis, seed suppliers can produce seed with hybrid vigour more cheaply while still protecting the investment. Apomixis is a system of vegetative propagation through seed, which occurs naturally in certain grass species. This has attracted interest of both public and private researchers because of new opportunities in breeding that transfer of this characteristic to major food crops (initially cereals) would offer in terms of using hybrid vigour in stable varieties. V-GURT-protection of apomicts has the best of both worlds: cheap seed production and an effective protection.

A more long-term benefit for certain breeders is the possibility to shield the use by competing breeders of particular genepools from use as parent material in further breeding. Currently, released varieties are widely used by breeders in order. Wheat breeding, for example, is largely based on the crossing of the two best varieties for the target area that are in the market at any time. Sometimes 'new blood' is added when disease resistances have to be introduced into elite materials, but commercial breeders do not have the financial capabilities to invest heavily in such breeding. GURT offer protection of this investment since released varieties can not be used for further breeding by competitors, who will have to work hard themselves to introduce such important traits into an acceptable genetic background.

Interests of Society

GURTs can be used for the environmental containment of transgenic seed (V-GURT) or transgenes (T-GURT). Where the GURT characteristic behaves like a dominant gene, outcrossing of a transgenic GURT plant with wild or local germplasm will not result in viable seed. GURT thus reduces the environmental risks associated with the introduction of transgenic crops. This advantage will be particularly important for the release of transgenics in the centres of diversity of the crop species.

A second interest of society is the increased investment in breeding that the technology is likely to trigger. Research investment in most major food crops is far below optimum levels. Public initiatives have tried hard to reduce this gap, such as breeding programmes in universities and public research institutes in the industrialised countries from the late 19th century onwards. Global public initiatives have initiated the Green Revolution and still encompass the main investments in breeding of major food crops that attract little private investments, like wheat and barley (CIMMYT, ICARDA), cowpea (IITA), bean (CIAT), chickpea (ICRISAT, ICARDA) and groundnut (ICRISAT, IITA). At the same time, private research in tropical maize, pearl millet and vegetables has been triggered by commercial opportunities, largely through the use of hybrids. GURT is likely to turn more crops into commercially interesting seed products, thus relieving some strain on the public research system.

Farmer's Interest

It can be in the farmer's interest to restrict the expression of a trait to a specific phase in the development of the plants. T-GURTs would enable a producer to restrict expression of a trait at will. Beyond that farmers may have limited direct interests in GURT-protected seeds. They may, however, benefit from the increased private investments in research when this leads to better varieties for their conditions. When GURT protected seed contains valuable characteristics they may take the GURT aspect for granted.

CONCERNS ASSOCIATED WITH THE INTRODUCTION OF 'GURT'

Since GURT is still in the development phase, actual effects cannot be determined. The potential effects can however be analysed when it is assumed that 'fool proof' GURT-protected varieties will be developed, which is likely to be technically feasible anywhere between 2005 and 2010.

Effects on Breeding

The biological protection that GURT offers will create novel commercial opportunities, especially in self-fertilising crops that are currently under-invested in. The breeding effort will however be directed towards commercial seed markets, i.e., to the higher-intensity farming systems.

Currently, less endowed farmers have access to plant materials from the formal sector (public and private) through so-called lateral spread. Poorer farmers may obtain some modern variety seed from neighbours and relatives that they may further multiply when the variety proves useful, either in a pure stand or when introduced in their diverse landraces. Supporting such lateral spread is a very effective means to spread the advances of breeding to resource-poor farmers and to remote areas. It has been official policy in many countries during the Green Revolution, and it is a main objective in modern participatory breeding and participatory variety selection programmes (Witcombe, 2001; Sperling & Ashby, 2000).

The widespread use of GURT in breeding is likely to increase the technology gap between the commercial farmers and those in less benign conditions. Breeding for the latter groups will have to be done by the public sector; modern varieties will not be available further adaptation to local conditions, either by farmers themselves or as part of participatory breeding initiatives. Also international public initiatives to support breeding for the less endowed may face problems to access new technologies and characteristics (e.g., disease resistances). Institutes like IRRI can access new biotechnologies for their rice research free-of-charge or at preferential conditions. When rice becomes a very commercial seed crop, however, they will be regarded a competitor by the commercial breeders.

When GURT is commercialised by the large multinational companies, the increased returns from seed sales may be used primarily to match the shareholder value that the same companies obtain in their pharmaceutics branches. In such case the increased research investment may be considerably less that envisaged and the farmers' interests may be limited accordingly. Such strategy is likely only when competition can be effectively excluded.

Effects on Seed Production

GURT offers very significant advantages for commercial seed production, especially in self-fertilising crops. Seed producers currently face strong competition from farm-saved seed, which can effectively be overcome through this technology. GURT furthermore allows for a cheaper 'hybrid' seed production.

The widespread use of GURTs is, however, likely to increase the gap between the 'larger' and 'smaller' seed companies, especially when large life science companies hold the key patents on the technology. These smaller companies may not be able to enter the new commercial seed markets created by GURT. Since their breeders depend more on the use of released varieties as parents in their breeding programmes than larger ones having their own genebanks and pre-breeding programmes, they are likely to be pushed to low value niche markets.

GURTs thus likely further strengthen the current trend of concentration in the global seed sector.

Farmers who produce their non-GURT crops adjacent to large areas of V-GURT fields of the same crop will face viability problems when using their own seed. In self-fertilising crops like most cereals, pulses and cotton, cross fertilisation rarely exceeds 2%, and viability losses will be negligible. Such minor reductions in seed viability are commonly compensated for by increased numbers of ears per plant in cereals and by increased leaf area per plant in legumes.

Introgression in truly cross-fertilising species like maize and oilseed rape may, however, go well beyond 20%, when small fields of local crops are surrounded by large areas of GURT crops. This will have a significant negative effect on crop yields. In wind-pollinated species like maize, a distance of 200 meters between GURT and non-GURT crops is sufficient to reduce the risks, but in insect-pollinated crops like sunflower and canola risks are considerable even at larger distances.

Effects on the Environment

Formal plant breeding leads to the development of uniform varieties that can be used by many farmers. Where GURT-protected (uniform) varieties replace genetically diverse landraces, a genetic erosion in the farmers' fields will be the result that is comparable to the introduction of Green Revolution wheat and rice varieties in India and Pakistan. If the additional plant breeding is focused on farmers who already use modern varieties (e.g., from public breeding), effects on crop genetic diversity will be minimal or even positive, i.e., when increased investments allow the breeders to use a much wider genepool or develop more varieties.

GURT may have a very positive effect on the risks related to the introduction of transgenic crops, especially in centres of diversity of these crops, as noted above. GURT can effectively contain geneflow from transgenic crops to such local germplasm or wild relatives, thus solving one of the major objections against genetic modification. GURT is therefore likely to increase the acceptability of GMOs in those areas where environmental concerns prevail. GURT on the other hand triggers a tremendous opposition in countries where the socio-economic aspects of genetic modifications prevail, such as in India.

Dependence and Seed Security

Farmers using GURT technology become completely dependent on seed suppliers. This may be comparable to hybrid seed users, except that in extreme cases these latter farmers have the option to use F2-seed. Farmers in relatively low-value markets in developing countries (i.e., value relative to other options for multinational seed companies) risk a lack of seed after poor seed harvests. Intensive horticulture producers already depend on the input suppliers and the introduction of GURT will not have very significant effects on these farmers.

Seed security is vital for all farmers. A highly competitive seed market guarantees seed security in industrialised countries. In many developing countries on the other hand, only one national or multinational company is active. Dependence on such monopolists is dangerous if farmers do not have local alternatives to purchase seed, especially at the lower side of the market. Multinational companies will supply their higher value market first when shortages occur due to ecological limitations or social unrest in the production areas. GURTs will increase the dependence on off-farm seed sources, thus creating a risk for the poorer farmers.

More pronounced seed security risks can be expected for the already seed insecure poor farmers who are not able to save their own seed every season. Risks of crop losses due to low viability will occur with the poorest farmers who depend on the grain market for their seed (often over 20% of farmers). They purchase something to plant at the last moment and risk unknowingly plant non-germinating (V-GURT) seeds.

This may also happen when food-aid is distributed to disaster-struck communities. Food grain is currently distributed as seed by ignorant relief agencies. Also, relief food supplies are often used as seed. Such disaster struck farmers may loose their investment in land preparation and loose a season's crop when GURT-food or 'seed' is supplied. It is the poorest farmers who risk loosing their crop this way. It may be argued that even the poorest will learn to test their seed before planting, but this will happen only after a number of them have had to learn the hard way.

POLICY CONSIDERATIONS

It may be clear from the above that the technology has some advantages and disadvantages. Policy makers basically have three options: to promote, to regulate or to prohibit the technology.

Even when policy makers do have a clear answer, it may not always be easy to identify the appropriate mechanisms to prohibit the development or the application of GURT.

The complexity of the issues also creates an institutional problem. Different ministries may consider different aspects.

National economic advantages of an increased national agricultural output through increased research investments for high potential farming systems have to be weighed against a possible increased technology gap between commercial and resource-poor farmers. In other words, lower urban food prices may coincide with increased food insecurity in remote rural areas.

Policy makers dealing with environmental issues have to take into account the risks of reduced agro-biodiversity on the one hand and reduced risks of gene-transfer from modern (transgenic) varieties to nature and endemic crop varieties.

Those interested in commerce may welcome an increased interest by multinational companies in national markets, but these advantages have to be weighed against possible reduced opportunities to develop a national commercial seed industry and against an increased dependence of farmers on foreign (owned) seed suppliers.

In the international discussion on V-GURT ('Terminator'), an important additional argument has been put forward very strongly: the point to develop sterile seeds is considered a major ethical issue. As an ethical issue this argument is close to the development of male sterility, which is common practice in hybrid production in crops like sunflower. More extreme cases in which the ethical argument has not been voiced are sterile triploid varieties (e.g., sugar beet), and seedless watermelon and grape.

The result of such deliberations is either to promote the use of such technology, to regulate them to particular uses, or to ban them from the country. Different countries may want to take different positions.

Promoting the Technology

Promotion of a new technology is commonly done through the granting of protection and subsidies. Subsidies seem unnecessary unless monopolies need to be avoided through the development of parallel publicly owned GURTs, but this is beyond the possibilities of most developing countries. Protection of GURTs through patents is possible, and has been granted in a wide range of countries already.

Prohibiting the Technology

It is not always very easy to prohibit a technology. Mechanisms in existing legislation have to be sought, because specific regulation to prohibit a particular technology is likely interpreted as a trade barrier and as such condemned by WTO.

The following options may be investigated.

Intellectual Property Rights (IPR) Laws

IPR legislation stipulates that all inventions that are novel, non-obvious and have an industrial application (patents), or that are distinct, uniform, stable and new (Plant Breeders' Rights) can be protected. GURT-based varieties are likely to be protectable in those countries that offer such protection systems. When GURTs appear to seriously conflict with food security of particular groups in society, the 'ordre public' clause in the TRIPs Agreement (Art. 27.2) may be used to outright disapprove protection. Using IPR law to ban GURTs would most probably, however, require a special provision to be added to the legislation, which would likely be the subject of a dispute in the WTO. It may, however, be considered to investigate the desirability and feasibility of adapting existing patent legislation, to avoid in international consensus undesirable impacts of GURT applications.

Moreover, disapproving protection on GURT does not necessarily mean that the technology will not be introduced in a particular country. This is particularly true in the case of GURT that offers a significant level of biological protection.

Biosafety Legislation

All GURT systems under development are genetically modified organisms. GURT products therefore are governed by biosafety regulations that prescribe certain procedures before they can be released into the environment (environmental safety), or when they are to be used in food production (food-safety). Such regulations are not developed to scrutinise the ethical or other objections against a certain technology. V-GURTs systems in insect-pollinated crops may be considered not environmentally safe because of the risks of viability reduction in farm saved seed in neighbouring fields, but the advantages of GURT restricting gene transfer into the environment will definitely be considered superior.

Similar to IPR, biosafety laws cannot be easily used to avoid the introduction of GURT when the technology itself does not offer particular environmental or food risks.

Seed Legislation

Conventional seed laws can be used to ban GURTs when they include a compulsory variety testing for value for cultivation and use (VCU). The Variety Release Committee may decide that the inability to reproduce seed (V-GURTs) or to reproduce vital values in the variety (T-GURTs) poses a serious value-reduction, thus turning the variety unacceptable. This must be considered a rather political move by the committee, especially when hybrids have not been subject to any such deliberations. Secondly, many countries have dispensed with this type of compulsory VCU-testing or maintained it only for certain crops.

Regulating the Technology

Existing or new regulatory instruments may address important concerns that surround the technology.

Concern

Small enterprises and public breeding programmes will face problems accessing new characteristics (e.g., disease resistances) when these are protected by (V or T) GURT.

Option

Similar to Intellectual Property Rights, where a temporary protection is granted in return for publication, a regulator may wish to oblige the GURT-owner to release non-GURT-protected variants with the same important characteristics available for further breeding to both public and private breeders after a period of grace–say 5 years. In case a breeder will need several years of conventional breeding to introduce the characteristic into his new variety, the GURT-owner will have had enough time to get his return on investment. Such a regulation would return a kind of breeder's exemption that will stimulate competition in breeding.

Concern

The commercial breeding effort triggered by GURT will be directed towards commercial seed markets. Less endowed farmers loose access to new materials through lateral spread. The widespread use of GURT in breeding will increase the technology gap between the commercial farmers and the less endowed ones.

Option

The main option seems implying a strengthening of public agricultural research and readjustment of its focus to those sectors that do not benefit from the commercial seed companies. This should mitigate negative consequences for the welfare of resource-poor farmers. This option contradicts the current trend of reduced public expenditure and privatisation, and it assumes that sufficient benefits can be generated for good researchers to choose for employment in the public service rather than loosing all the top-scientists to the private sector. This entails both national and international public research.

Concern

Farmers who produce their non-GURT crops adjacent to large areas of V-GURT fields of the same crop will face viability problems when using their own seed, especially in insect pollinated crops like sunflower and canola.

Option

The rule that the 'polluter pays' may be applied here. The farmer using GURT may have to observe a certain isolation distance from neighbouring fields from which seed will be harvested. He may however strike a deal with his neighbours, compensating them for the probable viability losses. A regulator may require such neighbours to have a written agreement signed before the season. In many countries however, such an option is likely not feasible. In areas of a country where an important genetic diversity of the crop (and wild relatives) are maintained, the use of GURT in such cross-fertilising species may be prohibited when their use creates problems to maintain local varieties. The question remains then whether such more commercial farmers should be compensated–this question relates to other forms of 'museum-farmers' to conserve agro-biodiversity, which proves unsustainable in most cases in the long run.

Concern

Formal plant breeding leads to the development of uniform varieties that can be used by many farmers. Where GURT-protected (uniform) varieties replace genetically diverse landraces, genetic erosion in the farmers' fields will be the result.

Option

When uniform varieties are better for farmers than their diverse landraces, it will not be possible to force them to produce landrace varieties. The main option is then to conserve the diversity in genebanks. Diversity can be supported, however through well-targeted programmes of breeding for diversity (Cooper et al., 2001). This requires a significant public investment (see above).

Concern

Farmers using GURT technology become completely dependent on seed suppliers, which is particularly risky in monopolistic seed markets. GURTs may well reinforce the concentration and integration trends in the seed industry and invite misuse of monopoly power.

Option

Very important in the current trends of concentration in the seed industry are effective antitrust legislation and antitrust institutions in developing countries and at the international level. This concern also warrants an active stimulus to create small-scale seed enterprises. Antitrust regulations may be very difficult to implement, especially in small developing countries. Experiences in seed enterprise development in such countries, on the other hand, are increasing (Kugbei et al., 2000).

Concern

Pronounced seed security risks can be expected for poor farmers who depend on the grain market for a large share of their seed needs, or those who are forced to use food-aid as seed.

Option

Where such risks are significant it may be necessary to allow GURT-protected materials only in closed chains, i.e., where the seed is supplied to farmers with a guarantee that all the produce is collected by one buyer, who works under a certification system that guarantees proper labelling. A less elaborate system may be to release GURT-varieties for use in particular geographical areas within the country where commercial farming prevails and where the produce is not likely to reach the grain market for (near) subsistence farmers.

Finally, relief agencies have to be well aware of the dangers of GURT-protected food aid. They should either make sure that they supply conventional food, or they should offer both emergency food and emergency seed supplies in well-marked packages. Even in the latter case it is strongly advised to refrain from food grain that may be GURT-protected.

CONCLUSION

Technological developments can raise opportunities and concerns that warrant policy response. Concerns may deal with various fields of expertise and different ministries may have to be involved in the analysis. The GURTs case illustrates that a technological development is likely to have a significant impact on a variety of issues, such as concentration in the seed industry, the public research system, the technology gap between commercial and marginal farmers, agro-biodiversity, etc. Opportunities to cope with the concerns fall within the competence of various organisations, which may be assisted or restricted by international treaties.

A close contact between researchers and policy makers is needed for a timely identification of scientific developments that may have considerable effects on society. Also the subsequent analysis requires input from both technical and social scientists. International organisations, both intergovernmental and non-governmental, can play an important role in identification and analysis. It has to be stressed though that policies have to respond to local situations and a balancing of the opportunities and concerns at the national level, using instruments that are available in national law and institutions.

REFERENCES

Cooper, H.D., C. Spillane & T. Hodgekin 2001. *Broadening the Genetic Base of Crop Production.* Wallingford, UK, CABI and Rome, FAO and IPGRI, 452 p.

Kugbei, S., M. Turner & P. Witthaut (eds.) (2000). *Finance and Management of Small-Scale Seed Enterprises.* ICARDA, Aleppo, Syria.

Kuvshinov, V. et al. (2001). Molecular control of transgene escape from genetically modified plants. *Plant Science.* 160: 517-522.

Sperling, L., J. Ashby, E. Weltzien, M. Smith & S. McGuire, 2001. Base-broadening for client-oriented impact: insights drawn from participatory plant breeding field experience. In: H.D. Cooper, C. Spillane & T. Hodgekin (eds.). *Broadening the Genetic Based of Crop Production.* Wallingford & Rome, CABI, IPGRI and FAO. pp. 419-435.

Visser, B., D. Eaton, N. Louwaars & I. van der Meer, 2001. *Potential Impacts of Genetic Use Restriction Technologies (GURTs) on Agrobiodiversity and Agricultural Production Systems.* Rome, FAO.

Witcombe, J.R., 2001. The impact of decentralised and participatory plant breeding on the genetic base of crops. In: H.D. Cooper, C. Spillane & T. Hodgekin (eds.). *Broadening the Genetic Based of Crop Production.* Wallingford & Rome, CABI, IPGRI and FAO. pp. 407-418.

Zuo, J. & Chua, N.H. (2000). Chemical-inducible systems for regulated expression of plant genes. *Current Opinion in Biotechnology* 11: 146-151.

SEED LEGISLATION
AND COUNTRY CASES

Seed Regulatory Reform:
An Overview

Robert Tripp

SUMMARY. There is general agreement that a national seed regulatory regime should respond to economic, political and technological factors specific to the particular country, but there is considerable controversy regarding the direction of regulatory reform. Regulation can be seen as a response to deficiencies in information. In the case of seed regulation, the major concerns are ensuring that farmers have adequate information about the seed that they purchase and that society is protected from negative externalities. Although these aims are clear, the performance of seed regulation is often problematic in terms of efficiency, relevance and transparency. Any approach to regulatory reform must acknowledge that it will be the outcome of political compromise; assign responsibilities for the distinct elements of regulation (standards, monitoring and enforce-

Robert Tripp is affiilated with Overseas Development Institute, London, UK.
Address correspondence to: Robert Tripp, Overseas Development Institute, 111 Westminster Bridge Road, London, SE1 7JD, UK.

[Haworth co-indexing entry note]: "Seed Regulatory Reform: An Overview." Tripp, Robert. Co-published simultaneously in *Journal of New Seeds* (Food Products Press, an imprint of The Haworth Press, Inc.) Vol. 4, No. 1/2, 2002, pp. 103-115; and: *Seed Policy, Legislation and Law: Widening a Narrow Focus* (ed: Niels P. Louwaars) Food Products Press, an imprint of The Haworth Press, Inc., 2002, pp. 103-115. Single or multiple copies of this article are available for a fee from The Haworth Document Delivery Service [1-800-HAWORTH 9:00 a.m. - 5:00 p.m. (EST). E-mail address: getinfo@haworthpressinc.com].

ment); and take full advantage of market mechanisms for transmitting information. *[Article copies available for a fee from The Haworth Document Delivery Service: 1-800-HAWORTH. E-mail address: <getinfo@haworthpressinc. com> Website: <http://www.HaworthPress.com> © 2002 by The Haworth Press, Inc. All rights reserved.]*

KEYWORDS. Regulation, regulatory reform

INTRODUCTION

Regulation is a subject that raises mixed emotions and contradictory interpretations in discussions of economic policy. A proposal for regulation may convey the image of steady, expert guidance, while at other times it may be seen as unwarranted meddling and the cause of needless distortion. These ambiguous perceptions are characteristic of debates about seed regulation, and they help explain why seed regulatory reform is a controversial element of agricultural liberalization policies.

Seed regulation encompasses two basic areas: seed quality control (certification and physical quality) and variety control (including registration, performance testing and approval). It is generally agreed that the most appropriate seed regulatory regime for a particular country will depend on specific economic, political and technical characteristics, and that as these conditions change so too should the regulatory regime. There is also general agreement about the direction of seed system evolution, in which the public sector usually plays an important initiating role in plant breeding and seed production while the private sector develops to eventually assume most of these responsibilities (Pray and Ramaswami, 1991). Despite this agreement, there is still considerable debate about recommended strategies for regulatory reform in developing countries that are undergoing liberalization. The fact that seed regulatory regimes in industrialized countries present a wide range of possible models makes the decisions even more difficult.

This chapter attempts to outline some of the most important elements in the debate. It begins by briefly reviewing the rationale for regulation. It then outlines some of the major pressures for regulatory reform. The following section looks at various strategies for instituting regulatory reform. The concluding section outlines some general principles for seed regulatory reform.

REGULATION

It is useful to see regulation as a response to deficiencies in the availability of information. In any modern economy, markets provide buyers and sellers with considerable information on which they can base their respective actions. But the flow of information is usually imperfect, and certain regulations may be imposed to address these problems. Similarly, a country's legal system provides guidance for the maintenance of public welfare and the establishment of equitable markets. But the legal system may operate at too general a level to address specific needs such as the monitoring and oversight of complex activities, calling forth the special boards or agencies associated with regulation (Friedman, 1985). In these cases a third party, often a government regulatory agency, steps in to help ensure that adequate information is available to guide market transactions or to enforce standards of public safety.

In the case of seed regulation, the major concerns are the provision of information to farmers (for their own economic well-being) and the management of information to control negative externalities in farming (ensuring that one farmer's actions do not jeopardize the welfare of others).

Lack of adequate information may affect farmers' abilities to make the best choices when they acquire seed. Several types of variety regulation are aimed at enhancing the quality of information about varieties on offer. Regulations may simply require that a variety is registered, or they may include specific performance tests to screen out unsuitable varieties. Variety registration frequently emerged in Europe and North America in response to confusion over variety names (and a profusion of synonyms) as commercial seed markets developed in the early twentieth century. In many European countries, mandatory performance testing was also established, leading to the present EU requirement for all field crop varieties to pass tests in at least one member country before being eligible for sale. In many developing countries, variety registration and performance testing were established as part of the public plant breeding system that virtually monopolized variety development until recently.

Although variety registration was originally designed to help farmers acquire more information from seed providers, it has also become an important tool for plant breeders and seed companies. In order to establish legal protection for their varieties, the procedures of plant variety protection (PVP) require some type of variety registration. Because of the large number of varieties now available, registration for PVP demands a considerable volume of very precise data on a variety's characteristics.

Even if farmers are well informed about the genetic characteristics of varieties on offer, seed is also a physical input that presents other information problems which regulation may address. Seed certification is a system that ensures that the purchased seed is indeed the variety it purports to be. In addition, seed certification is usually accompanied by tests for physical qualities that the purchaser may be unable to assess, such as germination potential and purity.

In addition to providing information to farmers, regulation often is designed to control environmental or economic externalities caused by farmers' choice of variety or seed. Farmers are unlikely to have access to this type of information, and even when they do there is no guarantee that they will necessarily act to protect the welfare of their neighbors. Variety regulation may prohibit the sale of varieties that are susceptible to diseases that can lead to epidemics. Seed quality control can guard against the introduction of weed seeds or diseased seeds that could cause widespread damage.

The advent of biotechnology has seen the emergence of entirely new areas of biosafety regulation, in response to the potential problems of transgenic varieties crossing with conventional varieties or with wild relatives. There is also concern that transgenic crops which incorporate insecticidal properties or viral components may affect other elements of the ecosystem. In addition, transgenic varieties may come under scrutiny from food safety authorities. Because transgenic varieties are subject to a range of performance issues as well as environmental and public health externalities, one of the greatest challenges is establishing which agency should take the lead in their regulation (Traynor, 1999).

Regulations are occasionally devised for purely economic motives as well. For many years Canadian authorities strictly controlled the types of wheat that farmers could grow in order to maintain the quality standard of exports, in the belief that this was the best way to enhance farmer incomes (Ulrich et al., 1987). In response to farmer pressure, several US states have enacted mandatory seed certification for potatoes in order to control the spread of particular diseases and thus to enhance the reputation of that state's potato crop (Makus et al., 1992).

Although it is clear that these instances of regulation attempt to respond to problems in the management of information, the actual performance of regulatory mechanisms often leaves much to be desired. The following section examines some of the major problems and possible responses.

THE RATIONALE FOR REGULATORY REFORM

The preceding discussion should make clear that the establishment and refinement of regulatory frameworks is a continual process. The direction of the process should depend on the institutional, technological and economic conditions under which regulation is to be conducted.

There are many concerns about the performance of regulatory institutions. Because regulation occupies a special position, above the market and beyond conventional law, it is not surprising that its isolation and privilege may lead to problems. An oversimplified, but nevertheless instructive, summary of the contributions and problems of regulation in general is provided by Bernstein's (1955) history of several US regulatory agencies. He proposes that although the agencies have very different mandates, all of their histories can be understood in terms of a common life cycle that begins with their gestation in response to public pressure; proceeds to a youthful, crusading stage; settles into a bureaucratic and routinized maturity; and ends in an ineffective old age, often including the agency's "capture" by the regulated industry. Regulatory capture is a familiar theme, and is explained by the fact that regulatory agencies depend on the industry for much of their information. Over time, regulator and regulated develop a comfortable relationship, with the industry often gaining the upper hand. Indeed, some analysts argue that much regulation can be interpreted as the outcome of pressure from the industry to protect itself from outside competition (Stigler, 1971).

Problems with seed regulatory performance are thoroughly described elsewhere (Tripp, 1997), but we can summarize the major difficulties in terms of efficiency, relevance and openness.

Someone, either farmers or the general public, must pay for regulation. The costs of a well-managed seed certification program account for a very small percentage of seed price, but in some developing countries the costs of variety registration and testing may be sufficiently high to discourage plant breeders or companies from entering the market. In many developing countries, variety testing and seed certification are supported by government budgets that are wholly inadequate. This means that varieties are slow to be tested and released and that much less seed is available because of inadequate resources for its certification. The establishment of PVP in developing countries will cause additional problems because of the high costs of registration and enforcement that must be faced.

The relevance of regulations to farmers' circumstances is also an issue. Variety performance testing can only be carried out in a limited number of sites, and their conditions may not be representative of those faced by the majority of resource-poor farmers. Seed certification standards may also be unreasonably high, thereby reducing the availability of adequate seed. A particular problem in developing countries is that the vast majority of resources for seed quality regulation is invested in the certification of seed production, even though problems are more likely to occur in seed storage or marketing, which require point-of-sale inspection.

In addition, a lack of transparency in the regulatory process is cause for concern. If varieties must be approved for sale, the composition of the board or panel that makes these decisions is crucial; in many cases such public entities are biased against private or foreign entrants. In addition, seed regulation is no different from many other types of regulation in presenting opportunities for collusion (where bribery can obtain approval for sub-standard products) or rent seeking (where regulatory officials demand extra payments to perform their duties).

A combination of restrictive seed regulations, inadequate budgets for maintaining the regulatory system, and a frequent lack of clarity about what regulations do or do not allow, is partly responsible for the lack of seed enterprise development in many countries. A recent study shows how these factors contribute to inadequate seed supply in several African countries (Tripp, 2000).

These problems with regulatory performance are exacerbated in the current climate of liberalization. Many public seed enterprises have been dismantled or privatized. In addition, there is a growing presence of private investment in plant breeding and private varieties are increasingly competing with those from public research institutes. Under such conditions it is very difficult to maintain a regulatory system that was developed when there was a single (public) supplier of agricultural technology and seed and that overlooks farmers' growing participation in agricultural input and output markets.

REGULATORY REFORM: THE SEARCH FOR ALTERNATIVES

This section considers some of the possible pathways for seed regulatory reform. It begins by reviewing the political dimensions of reform. It then delineates the various responsibilities that any regulatory system must fulfil. Finally, it considers the potential contributions of farmers and the seed industry in regulatory reform.

Political Factors

The term "seed regulatory reform" is often interpreted to mean the dismantling of mandatory government variety and seed regulation, but this discussion adopts a broader view. Discussions of regulatory reform must acknowledge the twin realities of imperfect information and imperfect regulatory remedies. Regulatory reform implies a search for the most effective and efficient means of transmitting information under a given set of conditions. Ideal solutions are unlikely, and the most appropriate compromises will be determined by a complex set of circumstances that are themselves subject to change. Regulation may sometimes be pictured as a neutral force that levels the playing field, but in fact it is usually the outcome of compromise among many political interests who are concerned with how the game is played. "[T]here are no optimal or best regulatory solutions, just solutions that respond better than others to the plural configurations of support and opposition at a particular moment in history" (Ayres and Braithwaite, 1992, p. 5).

Any effort at regulatory reform will involve balancing several competing interests. In developing countries, pressures from the seed industry and from donors pressing for liberalization are important factors. The realities of shrinking government budgets also threaten the capacity to maintain public regulatory agencies. The public agencies themselves are usually among the strongest defenders of the regulatory system, partly because of concerns for defending farmers' interests and partly because of instincts for bureaucratic self-preservation. In most cases, reform is not a comprehensive and carefully planned program but rather a series of changes and compromises affecting specific regulations and broader policies.

State control over seed provision in Peru broke down under the weight of government debt in the early 1990s. Mandatory seed certification was replaced by a voluntary system (which allowed considerable seed entrepreneurial activity to emerge), but nothing was done to reform strict government control over variety release and approval (Bentley, Tripp and de la Flor, 2000). In East Africa, a donor-sponsored initiative involved government representatives in a series of meetings that resulted in agreements to modify variety release requirements and to harmonize variety approval within the region, but with little change in the mandatory seed certification that all companies must face (Minde, 2000). India's seed policy reforms of the late 1980s allowed private companies greater access to public and foreign varieties. These policy reforms, combined with already flexible certification regulations, permitted a significant increase in private seed activity (Selvarajan, Joshi and O'Toole, 1999).

Although our discussion of regulatory reform focuses on instances where government regulatory responsibilities are reduced or re-directed, there are also instances where pressure for reform sees an increase in regulation. Perhaps the major contemporary example is the pressure for the establishment of intellectual property rights. The growth of private plant breeding and the emergence of biotechnology have provided strong incentives for the seed industry to press for plant variety protection (PVP). Nevertheless, there are still competing pressures. European countries were first to establish the Union for the Protection of new Varieties of Plants (UPOV) in the early 1960s; seed companies in the US were more ambivalent towards PVP at that time because they feared possible government regulation of registered varieties (Fowler, 1994). As PVP legislation is being debated in India, there is pressure from some parts of the government to use this to re-establish mandatory certification of the seed of registered varieties.

Regulatory Responsibilities

It is not useful to approach a particular regulatory regime as if it were an undifferentiated entity that should be defended, "reformed," or abandoned. Instead, it is useful to see regulation as composed of three elements (standards, monitoring and enforcement), each of which implies choices of responsibility and management. In any regulatory regime these three elements may be managed by the same, or different, agencies, which may be public or private. In addition, a regulatory regime may be mandatory or voluntary. Regulatory reform under liberalization offers opportunities for considering a wider range of management options for standards, monitoring and enforcement.

All examples of regulation are based on a set of standards. For instance, variety release may be dependent on performance in a specific series of tests. The standards may be determined by the same agency that monitors the tests (as is the case for many variety release committees in developing countries), or they may be standards that have been formulated with input from the seed industry or farmer organizations. The variety release standards may demand very specific levels of performance or they may simply be used to screen out obviously undesirable entrants (such as varieties likely to spread particular plant diseases). Similarly, seed certification standards may be those of the certifying agency itself or they may be standards that have been established by regional or international bodies in an attempt to achieve harmonization.

Regardless of how the standards are defined, a particular entity is assigned responsibility for monitoring compliance with those standards. For variety release, a government board or committee has usually been given responsibility for managing all of the testing, but where many of the varieties are developed by the private sector, companies themselves may be responsible for providing much of the data. In cases of regional release agreements, such as the EU's Common Catalogue of field crop varieties, decisions from one country are accepted by other member states without additional testing. Seed certification also presents several options. Certification may be completely managed by a government agency; the agency may license seed companies to do much of the testing while it merely carries out spot checks and supervision; or certification may be in the hands of private agencies (as in many states of the US).

Unless there is an effective mechanism for enforcement the regulatory regime will be of limited use. Enforcement may or may not be the responsibility of the same agency that does the monitoring. It is most helpful if the enforcing agency has a range of sanctions that it can apply, ranging from simple persuasion through penalties to license suspension (Ayres and Braithwaite, 1992). In India, seed quality monitoring is managed by state certification agencies, but enforcement is in the hands of the state departments of agriculture, who have the power to stop sales or, in extreme cases, temporarily close seed dealers or seed companies that are in violation of the regulations. In the US, monitoring is often done by private certification agencies but state seed laws are enforced by department of agriculture inspectors who are able to issue stop-sale orders or can request seizure of seed through civil court action (McDonald and Copeland, 1997).

The combination of standards, monitoring and enforcement by perhaps several different (public or private) agencies constitutes regulatory performance. In addition to decisions regarding the division of regulatory responsibilities, a particular regulation may be mandatory or voluntary. The strategy of making regulation voluntary is based on the assumption that those consumers (and companies) that are convinced of the value of regulation, and who are willing to pay for it, can have access to it. A potential weakness is the assumption that consumers have sufficient information to understand the costs and benefits of the particular regulation (Kelman, 1983).

The clearest contrast in regulatory strategies is between the US and the EU. In the former, there are no variety registration or performance requirements and seed certification is voluntary, while in the latter all field crop varieties are registered and tested, and seed must be certified. However, there is more that unites these systems than divides them (Tripp and Louwaars, 1997). In the US, varieties of many field crops are voluntarily submitted for evaluation to National Variety Review Boards, composed of representatives of government agencies and the seed industry. In both the EU and the US, the decision-making bodies have significant input from seed companies and also are subject to pressures from farming communities through democratic processes. In both cases the regulatory systems are the result of continuing debate and compromise between government, the seed industry and farmers.

Perhaps the area that requires most attention in seed regulatory reform in most developing countries is the need to strengthen the roles of the seed industry and consumers in the definition and management of seed regulation.

The Roles of the Public and Industry

Alternatives to formal third-party seed regulation require strengthening the capacities of farmers and seed providers to manage and interpret information. Hence regulatory reform should not simply be concerned with the adoption or abandonment of a particular regulatory protocol but rather should pay attention to strengthening the roles of all participants in the seed system so that markets and commercial law are more effective.

Seed regulation has often been established in response to farmers' concerns about their ability to recognize and maintain seed quality. In the early twentieth century, several state agricultural universities in the US helped set up certification programs as an educational tool for the farmers who were using public varieties and multiplying seed (Fitzgerald, 1990). Because of close interactions between the universities and the farmers these programs were successful and were the progenitors of several of the independent state certification agencies that are in place today. Farmers came to understand the purpose and value of seed certification. Studies in several developing countries have shown that most resource-poor farmers do not understand the meaning of the certification tags attached to the bags of seed that they buy (e.g., Tripp and Pal, 2000). Where farmers do not understand the certification process, regulation is less likely to have its desired effect, illustrating that even strict regulation is not a substitute for farmer education and empowerment.

Conventional consumer law may be an adequate alternative to seed quality regulation, assuming that farmers have access to, and an understanding of, consumer protection legislation. In developing countries, many of the seed quality problems that farmers face derive either from improper seed management by inexperienced dealers or deceptive practices (mislabeling, selling out-of-date stock) by merchants hoping to make a quick profit. Many of these problems are not detected and few result in sanctions, partly because the majority of seed regulatory resources are invested in the certification process rather than in point-of-sale inspection. In these situations, farmers' access to information about seed quality can be enhanced more by consumer education and strengthening consumer law, rather than by increasing "upstream" regulation.

Similarly, mandatory variety testing by public agencies should be compared to voluntary testing with strong participation from farmers. In developing countries, there are opportunities for public agencies or farmer groups or unions to organize testing programs. Increases in literacy and the growth of media (radio, television, farming publications) offer expanded options for providing information to farmers about the performance of new varieties. However, the effectiveness of voluntary testing depends crucially on farmers' political power and organization.

The identification of appropriate options for seed regulation also depends on the nature and organization of the seed industry. If the seed industry is young and inexperienced some type of third party oversight may be necessary (and welcomed by the companies) to guard against honest mistakes. As the industry matures its competence and reputation will increase, making outside technical supervision less necessary. One of the most important means for transmitting information about product quality is company reputation (Klein and Leffler, 1981). Companies invest considerable resources in brand names, advertising and company headquarters. These investments cannot be dismissed as mere promotional gimmicks; they transmit information to farmers about the companies' commitment and stability. In a competitive seed market it is not uncommon to find that companies are investing far more in quality control (often including independent testing) than would be required by any mandatory certification protocols. High value vegetable seed, which has almost never been subject to formal certification (because of expense), is sold mostly on the basis of company reputation.

Although company reputations and brand names can make a significant contribution to the flow of information in seed markets, these require time to become established. In developing countries, farmers may be uncertain of the identities of the companies that provide their seed. India has a well developed commercial seed market, but a study in two areas of Rajasthan showed that pearl millet growers often had trouble identifying the source of their purchased seed (Tripp and Pal, 2000). Many farmers could not distinguish between the state seed company and the many private companies that sold pearl millet seed. Farmers in the area that had longer experience with purchased inputs were somewhat more familiar with company names, but in both areas understanding of company names was strongly correlated with farmers' educational level.

Another option for controlling product quality and transmitting information is an industry association. Membership in the association is contingent on acceptable performance and the association can investigate any complaints from consumers. However, there are limits to the use of associations as a regulatory strategy, as company representatives may find it difficult to play policeman with fellow members unless the reputation of the industry is seriously jeopardized. In addition, industry associations can have contradictory effects. While they can establish standards and promote the industry, they (like other regulatory mechanisms) also can serve to lock out potential new competitors.

CONCLUSIONS

Seed regulatory reform should contribute to enhancing the quality and availability of information in order to improve the performance of seed markets. Conventional seed regulation is only one option for improving information flow. The degree of reliance on formal third-party regulation will depend on a number of factors. The availability of funding and personnel is important, because without adequate resources regulatory systems are ineffective and often counterproductive. The need for particular types of regulation also depends on the nature of a country's agriculture and the development of its economy. The choice of regulatory regime is rarely based solely on technical or budgetary considerations but is usually made in an environment of political struggle and compromise. In countries where public agricultural research and seed production have been predominant, the rise of the private sector requires a careful balance between ensuring that the public players do not maintain an unfair advantage but at the same time promoting responsible performance from the emerging private sector.

An assessment of regulatory options must include examination of the various components of regulation: establishing standards, monitoring compliance, and managing enforcement. Each of these three components may draw on different agencies (public or private, local or external). Where there are distinct types of demand for the information that a regulatory system provides, the system may be made voluntary. But whether voluntary or mandatory, seed regulation is most effective when farmers and the industry understand its purpose and operation.

There are many opportunities imaginative and well-focused seed regulatory reform. Most of these involve improving the flow of information between farmers and seed providers. They involve policies and programs well beyond the limits of formal regulatory frameworks. Emerging seed companies need to be encouraged to establish their reputations and to provide adequate information about their products. Input merchants may require training or supervision to improve their competence. Farmers require better access to information about the varieties on offer. They also need better organization as well as protection by the legal system in order to promote their rights as consumers.

Seed regulatory reform should not be seen as a choice between defending or dismantling a particular set of rules. It is better thought of as guidance and policy advice to encourage the continuing growth and evolution of national seed systems. It should draw on the resources and human capital of current seed systems and utilize these to take advantage of new opportunities.

REFERENCES

Ayres, I. and Braithwaite, J. (1992) *Responsive Regulation. Transcending the Deregulation Debate*. New York: Oxford University Press.

Bentley, J., Tripp, R. and de la Flor, R. (2000) 'Liberalization of Peru's formal seed sector,' unpublished paper.

Bernstein, M. (1955) *Regulating Business by Independent Commission*. Princeton, NJ: Princeton University Press.

Fitzgerald, D. (1990) *The Business of Breeding. Hybrid Corn in Illinois, 1890-1940*. Ithaca, NY: Cornell University Press.

Fowler, C. (1994) *Unnatural Selection. Technology, Politics, and Plant Evolution*. Yverdon, Switzerland: Gordon and Breach.

Friedman, L. (1985) 'On regulation and legal process,' in R. Noll (ed.) *Regulatory Policy and the Social Sciences*. Berkeley, CA: University of California Press.

Kelman, S. (1983) 'Regulation and paternalism,' in T.R. Machan and M.B. Johnson (eds.) *Rights and Regulation: Ethical, Political and Economic Issues*. Cambridge, MA: Ballinger Publishing Company.

Klein, B. and Leffler, K. (1981) 'The role of market forces in assuring contractual performance,' *Journal of Political Economy* 89:615-641.

McDonald, M. and Copeland, L. (1997) *Seed Production. Principles and Practices*. New York: Chapman and Hall.

Makus, L., Guenthner, J. and Lin, B.-H. (1992) 'Factors influencing producer support for a state mandatory seed law: An empirical analysis,' *Journal of Agricultural and Resource Economics* 17:286-293.

Minde, I. (2000) 'East Africa Harmonises Seed Policies,' *Agriforum* 13:4-6.

Pray, C. and Ramaswami, B. (1991) *A Framework for Seed Policy Analysis in Developing Countries*. Washington, DC: IFPRI.

Selvarajan, S., Joshi, D. and O'Toole, J. (1999) *The Indian Private Sector Seed Industry*. Manila: The Rockefeller Foundation.

Stigler, G. (1971) 'The theory of economic regulation,' *The Bell Journal of Economics and Management Science* 2:3-21.

Traynor, P. (1999) 'Biosafety management: Key to the environmentally responsible use of biotechnology,' in J. Cohen (ed.) *Managing Agricultural Biotechnology*. Wallingford, UK: CABI.

Tripp, R. (ed.) (1997) *New Seed and Old Laws. Regulatory Reform and the Diversification of National Seed Systems*. London: Intermediate Technology Publications.

Tripp, R. (2000) *Strategies for Seed System Development in Sub-Saharan Africa*. Working Paper series No. 2, Bulawayo: ICRISAT.

Tripp, R. and Louwaars, N. (1997) 'Seed regulation. Choices on the road to reform,' *Food Policy* 22:433-446.

Tripp, R. and Pal, S. (2000) 'Information and agricultural input markets. Pearl millet seed in Rajasthan,' *Journal of International Development* 12:133-144.

Ulrich, A., Furtan, W. and Schmitz, A. (1987) 'The cost of a licensing system regulation: An example from Canadian prairie agriculture,' *Journal of Political Economy* 95:160-178.

Seed Quality Control
in Developing Countries

Tony J. G. van Gastel

Bill R. Gregg

E. A. Asiedu

SUMMARY. Quality, in terms of purity and ability to establish a field stand of the desired plants, is the primary value factor of seed and a concern for every seed supplier. Thus, a primary task of the seed industry is achieving quality in production, maintaining quality in processing and handling, and establishing reproducible ways to measure quality and using these throughout the seed chain.

Seed certification and seed testing systems are aimed at providing high quality seed to farmers, and also at stimulating the seed industry, farmer use of better seed, and the national economy. The balance between internal (in-company) and external (official or private) seed quality control has to be adequate for the level of the country's seed industry development and farmer seed usage. Voluntary seed quality control, truth-in labeling and branding are alternatives to full seed certification systems that suffer from under-investment and under-staffing in many countries.

Tony J. G. van Gastel is affiliated with IITA/GTZ West Africa Seed Project, P.O. Box 9698, Accra, Ghana.

Bill R. Gregg is Consultant, P.O. Box 1756, Starkville, MS 39760 USA.

E. A. Asiedu is affiliated with Crops Research Institute, P.O. Box 3785, Kumasi, Ghana.

Address correspondence to: Tony J. G. van Gastel, ICARDA, P.O. Box 5466, Aleppo, Syria.

[Haworth co-indexing entry note]: "Seed Quality Control in Developing Countries." van Gastel, Tony J. G., Bill R. Gregg, and E. A. Asiedu. Co-published simultaneously in *Journal of New Seeds* (Food Products Press, an imprint of The Haworth Press, Inc.) Vol. 4, No. 1/2, 2002, pp. 117-130; and: *Seed Policy, Legislation and Law: Widening a Narrow Focus* (ed: Niels P. Louwaars) Food Products Press, an imprint of The Haworth Press, Inc., 2002, pp. 117-130. Single or multiple copies of this article are available for a fee from The Haworth Document Delivery Service [1-800-HAWORTH 9:00 a.m. - 5:00 p.m. (EST). E-mail address: getinfo@haworthpressinc.com].

Government policies and subsequent national legislation have to carefully balance the advantages and disadvantages of the different options. Governments can promote the development of a competitive seed industry through (compulsory or voluntary) seed quality control services while at the same time protecting their farmers. As a competitive private sector develops, the government's role may gradually decline. Despite the needs for a certain level of international harmonization, there is no single blueprint solution to the regulation of seed quality, nor are there standard solutions to the institutional aspects *[Article copies available for a fee from The Haworth Document Delivery Service: 1-800-HAWORTH. E-mail address: <getinfo@haworthpressinc.com> Website: <http://www.HaworthPress. com> © 2002 by The Haworth Press, Inc. All rights reserved.]*

KEYWORDS. Seeds, viability, purity, seed health, certification, quality assurance, internal quality control, branding, truth-in-labeling, fiscalization, quality declared

INTRODUCTION

Quality, in terms of purity and ability to establish a field stand of the desired plants, is the primary value factor of seed.

Thus, a primary task of the seed industry is achieving quality in production, maintaining quality in processing and handling, and establishing reproducible ways to measure quality and using these throughout the seed chain.

Over many years of seed industry development, seed quality control has evolved from simple visual observation to comprehensive measurements at all stages of production and marketing.

A seed enterprise varies from a single farmer who produces seed for local sale, to the international company with many employees and operations in many countries to breed varieties, produce and market many kinds of seed, over wide areas. In every seed enterprise, quality control is of major importance. Every operation must strive to produce–at the least cost and loss–the best quality seed that can be sold at prevailing market prices and market demand.

A quality-assurance system is not purely aimed at providing high quality seed to farmers. It also aims at stimulating the seed industry, farmer use of better seed, and the national economy. It must be realistic and adequate for the level of the country's seed industry development and farmer seed usage. To meet the existing needs, quality control (QC) systems in developing countries are different from those in developed economies.

This paper emphasizes quality control approaches for developing countries, and is mainly focused on sub-Saharan Africa.

INTERNAL QUALITY CONTROL

Internal Quality Control (IQC) is an internal operation within a seed enterprise, an essential part of the management of the company/enterprise. It is a voluntary approach to ensure that the seed of the company is of high quality. Its purpose is not only to supply quality seed to farmers, but also to ensure a high level of farmer trust in the enterprise, help develop and maintain markets, and prevent losses and excessive operating costs. The company decides on the sophistication of internal quality control operation, as well as the level of seed quality.

Every operation, from the informal farmer seed producer to the multinational corporation, should have some form of IQC. IQC ensures that the company's seed is of high quality (at least, in terms of the cost expended vs. price charged) and that it will attract farmer-customers and avoid problems which may alienate customers or even result in damage lawsuits.

In a typical developed-economy seed company, IQC is a "staff" function; it consists of one or more specialists who report directly to management. They essentially serve as the quality control arm of management. IQC is expected to inspect all operations, test seed at all steps, and stop/change anything to ensure that the desired quality and operating efficiency are reached.

In theory, only IQC is required to make sure that quality seed of improved varieties reaches farmers. This assumes that each seed-producing/marketing entity has an efficient and effective IQC, which operates independently from company 'economics.'

IQC is also a 'cheap option' in terms of government resources, because effective IQC means that only–if any–a simple external quality control system needs to be maintained by the government.

EXTERNAL QUALITY CONTROL

External Quality Control (EQC) is carried out by an independent agency, often governmental. EQC should ideally be completely independent of seed producers and suppliers. EQC does not answer to seed enterprise managers, but is based on independent tests and inspections. External seed quality control could be performed by a seed industry association–as is the basis of seed certification and control in The Netherlands and some states in the USA, but in most countries EQC is performed by government agencies. Some forms of EQC are made compulsory once they are implemented.

EQC establishes whether the laboratory and field standards set by the Seed Law and its regulations are met. In developing seed systems, the certification agency may develop its own standards in the absence of a formal seed law.

Such standards are commonly established in close collaboration with the (formal) seed producers.

The primary goal of EQC is to help develop the agricultural economy by assuring farmers of high-yielding seed quality. At the same time, it helps to develop and maintain a stable quality-oriented seed industry to supply the seed needed to produce the required food/feed/fiber crops. For the government, external quality control is demanding in terms of resources such as personnel and operating costs.

SEED LAW REGULATION

Under the authority of a National Seed Law, seed quality is regulated and must meet established minimum standards. These standards are often quite low (as compared to what would be possible under a strong quality control program), and are intended to prevent sale of very low quality, crop-damaging seed to farmers, so that farmers receive seed of basic levels of quality.

As a result of international emphasis, most countries, even those with developing agro-economies, have a Seed Law. Seed Law structure, requirements, staffing, and implementation vary widely, in response to the role of the government in the seed industry, the government's financial ability, local agroeconomic conditions, and local seed supply development.

In general, the Seed Law establishes authority for the government to control seed quality, and establishes a government agency to implement it. Regulations under the Seed Law then establish operating procedures and minimum seed standards. Procedures and minimum seed standards are nationally determined and may differ from country to country. However, to promote agricultural development through country-to-country exchange of seed, harmonization of Seed Laws is being achieved among European countries, as well as in some other groups of countries.

The Seed Law implementing agency operates with (1) inspectors who are legally authorized to enter seed enterprise premises and sample seed offered for sale; (2) official seed testing laboratories to test the seed; and (3) the legal power to stop sale of seed which do not meet standards.

Seed Law regulation is essential and generally protects farmers and promotes development of quality-oriented seed suppliers. However, it has weaknesses, including:

1. Required seed quality standards are often so low as to minimize the value-added effect of the seed.
2. Genetic (varietal) quality is quite important, but is not well controlled.

3. "Informal sector" seed sold farmer-to-farmer is often exempt from meeting requirements of the Seed Law. This is often a major part of the seed actually planted, so the Seed Law is only partially effective.
4. Implementation of the Seed Law is often weak and inadequate due to lack of funding and effective personnel.

Seed Testing

Seed are generally tested only for "planting" quality factors–germination, physical purity, and presence of undesirable materials such as inert matter, other crop seed or weed seed. Genetic or varietal purity is generally not involved, as this is usually difficult to determine from seed characteristics.

Seed testing is relatively easy, basically it involves checking how much seed of the relevant species is included in the seed lot, and checking whether the seed actually germinates. This can best be done just before planting and under the farmers' conditions. This is, however, hardly ever the case and the seed is tested long before it is planted at a location, which has no relation to farmers' conditions.

Germination potential will decrease with storage and germination conditions may be very different at the farm, especially when seeds are traded internationally. Also, seed borne diseases and weed seeds may be a small problem in one site and may be strongly regulated in other places.

Seed quality information is only of interest if it can be reproduced at any time at another location. The International Seed Testing Association has, therefore, developed standardized seed testing procedures which are the basis for uniform seed testing worldwide (ISTA, 1999).

For small companies or developing national seed systems it may be difficult to fully adhere to ISTA-rules. Seed testing results must, however, be reproducible, even if only locally available materials can be used (van der Burg 1999).

Certification

As breeding of more productive varieties progressed, it became obvious that germination and physical purity of seed were not enough to assure farmers of seed which carried the genetic potential for disease resistance, higher yields, and other crop factors. Thus, standard seed quality testing did not have full effect in increasing farmer profit and the society's supply of readily-available food.

Special programs called Certification systems were developed to ensure seed of higher-than-usual quality, including both genetic (varietal) purity as well as the other factors of seed quality. In developing countries, the certification implementing agency is often within the government; occasionally it is an independent unit.

To ensure varietal purity, Certification sets standards, requirements for

1. The seed that is used to plant seed production fields.
2. Land requirements.
3. Off-type plants, other varieties, seed-born diseases in seed fields.
4. Inspections of seed fields.
5. Harvesting, handling, cleaning, bagging, etc.
6. The number of generations in order to ensure varietal identity and genetically pure seed.

Seed testing is always a part of certification and the certification agencies takes samples (usually after processing) of each seed lot to be certified.

Certification records have to be complete to allow tracing any bag of seed back to the field which produced it, and from there back to the original breeder seed from which it descended.

Certified seed bags carry special tags, seals and lot numbers. Procedures and standards are nationally determined, so they differ from country to country. However, different groups of countries have formed associations to standardize procedures among all the member countries such as the Organization for Economic Co-Operation and Development (OECD, 1973 and later updates) and the different states of the USA. The OECD rules and label design has been adopted by many countries. This permits easy seed trade among the member countries/states.

Certified seed is usually the best seed available. However, it has several weaknesses:

1. The costs to implement the activities required to certify seed varietal identity and purity and to test all the quality aspects raise the selling price of the seed.
2. Because of the higher cost, fewer farmers are able or willing to purchase certified seed. Only the more progressive, financially strong farmers use them.
3. Breeders often do not provide adequate morphological/agronomic descriptions (or marker characteristics) of varieties, so it is difficult for field inspectors to identify lower-yielding contaminants or off-types.
4. Due to inadequately trained inspectors, insufficient operating funds, etc., fields are sometimes not adequately inspected or quality adequately maintained. This results in lower quality of certified seed, or higher losses of seed at the testing phase. In the first case the certification label

will not become a quality label, potentially destroying the complete national seed development effort.

The system including Seed Law regulation combined with certification is referred to as Comprehensive Regulatory Seed Quality Control. A good overview of such a comprehensive system is presented by Wellving et al. (1984).

Truth in Labeling

The Truth in Labeling system, used in many countries in the Americas, allows basically any seed to be sold as long as its quality is clearly stated. The producer/seller must make sure that the seed is tested and that the result of the test are indicated on the label. The buyer must decide whether he likes the quality of the seed based on the information provided on the label. The government's task is to set up a system that ensures that the information on the label is correct and complete, and to prosecute traders who market improperly labelled seed or seed with a lower quality than stated.

Advantages of this system are primarily that government is freed from the responsibility to ensure that farmers get seed of a minimum quality standard. This is a main reason for the promotion of this system in developing countries (Gisselquist & Pray, 1999). It also permits seed companies to market seed, which may be below the levels of what may otherwise have been minimum legal standards.

Disadvantages are that farmers in developing economies may not understand seed value and quality factors and that they are not accustomed to comparing seed and looking for quality. Illiterate farmers may consider any label as a quality guarantee irrespective of the printed data.

Moreover, due to the malign storage conditions in most tropical countries (high temperature and relative humidity), testing results that are several weeks old, may not be relevant at the retail level. Although this would also often be true under a certification system, but–at least in theory–the certified seed should be checked before it is marketed.

Fiscalization

The Fiscalization system, in operation in Brazil, markets "Fiscalized Seed" whose quality is assured by the company's agronomist, who is accredited by the government's Fiscalization Office (FAO Expert Consultation, 1986). Seed standards are set by the Fiscalization Office and the company. This system is clearly a mix of IQC (an agronomist of the company 'certifies' the seed) and EQC (Fiscalization Office accredits the agronomist). The accredited agronomist makes inspections and takes samples. Laboratory tests may be conducted by an official government lab or by an accredited private lab.

Properly implemented, this is an advanced, more farmer-friendly form of the truth-in-labeling approach, provided that the seed company is truly quality-oriented and its Fiscalization agronomist is dedicated and "bonded" to ensure that all standards are truly met. It requires more expenditure on IQC by the seed company, but less on the part of the government. However, total expenditure is less, and seed quality is improved more efficiently. It also works best where there is an active independent (government) Agricultural Extension program to promote farmer use of quality seed, and/or an agro-economy (such as Brazil's) where there are many large farms which operate as businesses.

It also requires a governmental EQC program, which makes random spot-checks, in addition to lab tests, to ensure that quality-assurance operations are carried out effectively.

Quality Declared

The Quality Declared system, developed by an FAO expert consultation (FAO, 1986), as far as the authors are aware, is used officially only in Afghanistan. Again, government sets the standards and seed is produced only of varieties that have passed official government-conducted tests. The system allows the producer to declare the seed quality, based on tests that the company has arranged to be conducted. The seed is then marketed as 'Quality Declared Seed.' The Quality Declared seed system gives major importance to IQC, since the company declares the quality of the seed.

This system was basically designed to provide governments with a relatively cheap option for quality control in developing countries. It is much cheaper than the Comprehensive Regulatory approach and thus requires less government resources. On the other hand, it still is expected to provide sufficient guarantee for farmers to receive quality seed.

To work effectively, the Quality Declared system requires seed companies with sufficient resources to maintain levels of internal quality control sufficient to achieve higher seed quality. A disadvantage is that in many developing agro-economies, most seed suppliers are small operations, which have difficulty in effectively conducting the required operations.

HOW SEED QUALITY CONTROL OPERATES
IN DEVELOPING COUNTRIES

In developed agro-economies, farmers generally produce for the market, are educated and have the financial ability to pay for higher-yielding seed. The seed industry is well-developed; financially strong seed companies offer different seed varieties, and have intensive internal quality control systems. Often, the company's brand name signifies to the farmer, a higher level of seed quality than a Certification tag. And, the farmer is free–and able–to make a wise selection. A Seed Law and Certification back up the seed industry's IQC efforts. Producers and buyers are all very quality conscious.

Brand Names

In general, farmers–not only educated business farmers in developed agri-cultures–look for something simple to associate with assured quality when they buy seed. Better seed companies sell seed on the basis of their brand name or "trade mark." A brand name identifies the seed of that specific company; to protect the integrity and farmer trust in that brand name, the company strives to sell only high-quality seed. Often, the company's internal standards are higher than standards set by government authorities.

A seed enterprise requires long-term investments, which in turn require that the enterprise maintain consistent year-to-year sales. To ensure that farmers become "repeat year-after-year buyers" of their brand name seed, the compa-nies carry out considerable marketing, advertising and promotion to farmers. The first promotional effort is to maintain consistently high seed quality, so that the farmer trusts the seed and always tries to buy it. Many seed enterprises realize the importance of quality in marketing, and normally absorb the loss of discarding substandard seed rather than risk damaging the company's reputa-tion by selling it for a "quick profit."

This is thus a strong and very effective voluntary quality control system. IQC is a very important part of the company's management and EQC plays a minor role. In fact, in developed countries the importance of EQC is declining.

Under competitive conditions (different companies operating in the same market), the quality of seed is high in order to secure a market for the com-pany's brand-name seed. Under such conditions, there is no longer a need for compulsory Certification.

Formal Seed Quality Rules

Despite the overall importance of brand names in seed marketing in com-petitive seed markets, most industrialized countries have arranged for some level of External Quality Control. In the European Union, production and sup-ply of field crop seeds is heavily regulated with compulsory variety registra-tion, seed certification and seed testing by official seed certification agencies. For vegetable seeds certification and quality control is done by the companies themselves under the responsibility of the certification agency. The agency prescribes the procedures and standards, and monitors the quality assurance system of the seed supplier rather than the seed itself. This is a strongly regu-lated IQC system.

These systems are generally supported by the industry.

OPTIONS FOR DEVELOPING SEED INDUSTRIES

A developing agro-economy is often characterized by most farmers having small landholdings, being poorly educated and weakly financed. Relatively few farmers are well-to-do, use quality inputs, and produce for the "market."

Most of these small farmers are uneducated, they do not understand the formal seed production technology nor the value of certified seed. Thus, they are unable and/or unwilling to pay higher prices for seed. Much of their production is often for home consumption or barter and they are little affected by the market for agricultural produce. There are few incentives to produce for the market and thus to invest cash in seeds and other inputs. Over 90% of agricultural production is derived from farmer-produced seed (Almekinders et al., 1994).

The market for certified seed is thus very small, and not many true "seed enterprises" operate successfully in the local seed supply-use situation. Where they function they are usually small, low-overhead operations. They often all sell the same product/varieties to the same market, so hardly any competition exists.

Brand Name

The Brand Name approach would–in theory–be the best option for staff and capital constrained situations, even for developing countries, because it would render all other quality assurance measures unnecessary so there would be no need for expensive government-implemented EQC. However, it cannot be effective unless it is coupled with:

a. seed enterprises which can implement strong, efficient, effective and reliable IQC;
b. enough seed enterprises to create true competition;
c. strong government extension promotion to farmers; and
d. enough farmers purchasing quality seed to create a true "market" which the seed enterprises can service.

A disadvantage is that the Brand Name approach only works when there is an open competition in a particular seed market, while in many developing seed industries there is little or no market and/or competition. Even if there are different supplyers, they tend to sell the same varieties. Furthermore, markets in developing countries are still small and consequently enterprises are small or very small and can not afford comprehensive IQC operations.

Also, IQC accept/reject decisions are entirely an internal company matter; they can be over-ridden by management in order to achieve a desired level of sales, avoid loss of seed, or other economic reasons. Thus, a system based heavily on IQC is not very appropriate for developing seed industries, as an assured level of quality may not be guaranteed. Under such conditions, it may be guaranteed only from larger seed companies that can absorb the loss of seed, which are below the set internal quality standard. Such companies hardly exist in developing agro-economies.

Therefore, the Brand Name system is effective only in developed seed industries where farmers are educated and have adequate financial strength to buy quality seed, and the "market" offers several kinds of seed so farmers have a choice of what to buy–and are sufficiently trained and educated so they can make a competent choice.

Comprehensive Seed Quality Control

Compulsory comprehensive external regulatory quality control could ensure that farmers receive the best quality seed. However, this option has been tried for many years, and has not been very successful. Government fails to see the need to develop seed (and often, to develop rural economies), so considers it as too expensive to implement effectively. Simultaneously, farmer education is not conducted effectively, and the seed quality control implementation is not uniform and seed quality is not as high–or dependable–as it should be. The result is a significantly higher selling price of seed that often has an only marginally higher quality. Farmers in developing countries often cannot afford or are not willing to pay this higher price. Thus, government quality control agencies are heavily subsidized, do not service a large part of the farming community, and thus are not sustainable.

Truth-in-Labeling

A Truth-in-Labeling system also does not seem to be appropriate for developing seed industries where farmers are not yet sufficiently educated to judge information provided on a label, there is little effective extension promotion of quality seed, and farmers seek "the lowest cost." However, a Truth-in-Labeling system can be much cheaper for government, and thus appears more adequate for developing countries than the comprehensive regulatory approach–if there are enough seed companies of sufficient financial resources to implement it.

Quality Declared Seed

The Quality Declared System should provide reasonable quality for a reasonable price to farmers–but again, if there are enough seed companies of sufficient financial resources to implement it.

LIBERALIZING AND HARMONIZING

For agriculture to be successful today in fulfilling its national role of (1) rural employment, (2) food production, and (3) market commodity production, a strong private seed sector is essential. But, a strong private seed sector can develop in developing agro-economies only if companies have easy access to the latest technology and economic support, and can market their seed across national borders. These conditions are essential to create a market adequate to support a seed industry.

Currently, it is argued that to achieve this requires deregulation, liberalization and harmonization of seed regulations. For instance, the World Bank's Sub-Saharan Africa Seed Initiative proposed a number of "best practices for seed regulations" to try to get the private sector interested in investing in sub-Saharan Africa. These "best practices" include:

- Entry of companies into a market for seed should be simplified. There should be no heavy requirements for investment or specific conditions for trained staff, facilities, etc.
- Introduction of new varieties should be possible without waiting for official tests and approvals. This is intended to simplify moving varieties/seed across national borders, so seed can be sold in several countries by one company.
- Voluntary variety registration should be allowed.
- A Truth-in-Labeling system should be implemented with regard to seed offered for sale.
- Import and export regulations should be simplified.
- Seed certification should be voluntary.

The authors support the idea of deregulation, liberalization and regional harmonization so that it will become easier for seed entrepreneurs to enter into the market and to operate more profitably. However, deregulation should not result in a situation where companies and entrepreneurs have all freedoms and rights; a certain degree of protection for farmers is necessary to ensure that they receive quality seed. Certainly in developing countries, there are some entrepreneurs who are after "quick money" and not willing to invest in activities which have a long-term profit horizon.

CONCLUSIONS

The first requirement is that government must realize that it is the only agency which can initiate, encourage and support rural development and seed suppliers.

In developing agro-economies, governments may not be rich, but they still have resources far superior to those of the small individual farmers who make up much of the rural population and the agriculture. Therefore, governments have an important role to play in developing and organizing rural economies. Governments must play a role in (1) maintaining strong extension farmer education, (2) creating incentives for private sector investment in rural enterprises such as seed supply, and (3) ensuring basic levels of seed quality.

As the private sector develops, and a seed market is created, the government's role may gradually decline.

A certain degree of control will be necessary in a market where competitive forces are not yet operating. In fact, the more competition, the less regulation is necessary. With regard to seed quality control, this means that at initial stages, probably some EQC will be required, but that–as more companies enter into the market–more and more reliance can be placed upon ICQ. EQC should have as its major task, to stimulate and ensure that companies are able to–and do–carry out IQC.

Governments can thus promote the development of a competitive seed industry while at the same time protecting their farmers. The careful balance between regulating and stimulating enterprise development calls for maximum transparency and participation in both the design and the implementation of rules for mandatory or voluntary seed certification and quality control (Tripp & van de Burg, 1997). Where international seed trade is important (either exporting or importing), it is important to harmonize procedures and standards as much as possible.

As identified above, the certified seed producers currently supply only a very small portion of the total quantities of seed used for most crops in developing countries. Also it is important to design regulations in such a way that they do not obstruct the existing local seed production and exchange system (Louwaars, 2000).

There is no single blueprint solution to the regulation of seed quality, nor are there standard solutions to the institutional aspects (certification agencies within government or under the management of seed associations). The optimal system depends on the general level of agricultural development in the country, the development level of a competitive national seed industry, the importance of the local seed systems. These aspects differ among countries, but also among different seeds in any country.

REFERENCES

Almekinders, C.J.M., N.P. Louwaars & G.H de Bruijn, 1994. Local seed systems and their importance for improved seed supply in developing countries. *Euphytica*, 78: 207-216.

FAO, 1986. *Quality declared seed*. Rome, FAO.

Gisselquist D. & J. Srivastava, 1997. *Easing the barriers to movement of plant varieties for agricultural development.* Washington, DC, World Bank Discussion Paper No. 367, 139 p.

Gisselquist, D. & C. Pray, 1999. *Deregulating technology transfer in agriculture: Reform's impact on Turkey in the 1980's.* Washington, DC, World Bank, 62 p.

ISTA, 1999. *Seed rules.* Zurich, International Seed Testing Association.

Louwaars, N.P., 2000. Seed regulations and local seed systems. *Biotechnology and Development Monitor* 42: 12-14.

OECD, 1973. *OECD schemes for the varietal certification and seed moving in international trade.* Paris, France, Organisation for Economic Coopperation and Development.

Tripp, R. & W.J. van der Burg, 1997. The conduct and reform of seed quality control. pp. 121-1569. In: R. Tripp (ed.). *New seed and old laws: Regulatory reform and the diversification of national seed systems.* London, Intermediate Technology Publications.

van der Burg, W.J., 1999. Seed quality testing. Pp. 147-159 in: C.J.M. Almekinders & N.P. Louwaars (eds.). *Farmers' seed production, new approaches and practices.* London, Intermediate Technology Publications.

Wellving, A.H.A., I. Kiliangile, P.M. Jones & D.M. Naik, 1984. *Seed production handbook of Zambia.* Lusaka, Department of Agriculture, 397 p.

Variety Controls

Niels P. Louwaars

SUMMARY. Variety registration and testing systems were developed in industrialised countries in the first half of the 19th century in order to create transparent seed markets. Differing seed policies in Europe and the USA, based on different perceptions of the role of the government in economic life created very different government involvement in the regulation and implementation of variety controls. Europe developed a system, based on public institutions, whereas the market parties largely remained responsible for the voluntary variety registration system in the USA. Most developing countries followed the European example when they developed their formal seed systems during the Green Revolution.

The model showed hard to implement during the first decades of their existence. Current variety controls are under pressure of the structural adjustment of the economy leading to a privatisation trend in the seed sector.

The present paper presents the objectives for variety control and the three main functions: variety registration, performance testing, and release decisions. We then discuss the limitations in the practical implementation, and current trends in restructuring variety controls.

It seems inevitable that countries have some kind of variety registration system that identifies varieties. There are various ways to making the performance testing of varieties more efficient and less obstructive to seed industry development. Participation, transparency and international cooperation are the keys. *[Article copies available for a fee from The Haworth Document Delivery Service: 1-800-HAWORTH. E-mail address: <getinfo@haworthpressinc.com> Website: <http://www.HaworthPress.com> © 2002 by The Haworth Press, Inc. All rights reserved.]*

Niels P. Louwaars is affiliated with the Plant Research International, P.O. Box 16, 6700 AA Wageningen, The Netherlands.

[Haworth co-indexing entry note]: "Variety Controls." Louwaars, Niels P. Co-published simultaneously in *Journal of New Seeds* (Food Products Press, an imprint of The Haworth Press, Inc.) Vol. 4, No. 1/2, 2002, pp. 131-142; and: *Seed Policy, Legislation and Law: Widening a Narrow Focus* (ed: Niels P. Louwaars) Food Products Press, an imprint of The Haworth Press, Inc., 2002, pp. 131-142. Single or multiple copies of this article are available for a fee from The Haworth Document Delivery Service [1-800-HAWORTH 9:00 a.m. - 5:00 p.m. (EST). E-mail address: getinfo@haworthpressinc.com].

KEYWORDS. Variety registration, variety release, performance-testing, DUS-testing

THE ORIGIN OF VARIETY CONTROLS
IN INDUSTRIALISED COUNTRIES

Variety controls are a common element of national seed policies and seed regulatory frameworks. Compulsory variety registration has evolved in Europe in the first half of the 20th century under the influence of the farmers associations and the seed industry itself. 'Bona fide' breeders and seed producers wanted to create transparency in the market where variety names and varieties themselves were used for commercial purposes only.

Clarity was lacking in both directions:

1. Each seed supplier gave varieties a new name in order to create a brand for the company. Claims were made to distinguish the product from the competitor's (supplying the same variety), that may have been valid when they dealt with seed quality aspects (purity, germination, etc.), but that were ungrounded when they dealt with for example specific adaptation or disease resistance.
2. When a particular variety became popular it was easy to re-name another variety with the same or a very similar name in order to increase sales.

The solution to curb such practices was a central variety registration system that linked one name to one variety. Such registration was based on morphological descriptions, complemented with some striking agricultural characteristics. The registration system was used in The Netherlands in the 1930s to design a potato breeders' remuneration system, based on the acreage planted to each variety. This created a basic revenue system for breeders that further evolved into Plant Variety Protection.

Testing for the value for cultivation and use (VCU) of varieties originated in the variety testing systems designed by farmers' associations in various countries in order to validate claims by the seed suppliers. Such associations offered prizes for the best varieties in agricultural fairs. This system created an incentive for breeders, both in terms of prize money and also as 'free advertisement' for their seed.

The first 'register of qualified varieties' was established by the German Agricultural Society in 1905 (Rutz, 1990). Such registers became compulsory when national seed laws were enacted in Europe in the 1940s. In the UK, a voluntary variety registration system remained until 1964.

The main objectives of the legislators were to create transparency in the seed market and to support the serious seed producers and reduce the competition from 'fly-by-night' suppliers. They furthermore intended to stimulate plant breeding. Similar local initiatives in the USA were not formalised at the government levels to the same extent in that country, even though differences exist between states. The seed market in the USA is considered transparent enough, the testing of varieties and publicising the results is considered a task of market players (seed suppliers and farmers themselves), and branding through variety names a good way to promote a competitive seed market. The result was, however, that there were few incentives for seed companies to develop their own varieties, and breeding of many major cereal and pulse crops remained a public or university task.

VARIETY CONTROLS IN DEVELOPING COUNTRIES

The historic development of formal seed production systems was very different in most developing countries. Seed production became a political focal point after the Green Revolution, when the results of the breeding programmes had to be taken to as many farmers as possible. Public seed multiplication programmes were set up with a strong support from the FAO Seed Industry Development Programme. Since there were few commercial incentives in the public seed production units to give sufficient emphasis on seed quality, special seed certification units were developed in this framework, similar to the institutional organisation in the North. Variety controls, particularly those based on multilocational testing of varieties, were part of this system, as they fitted very well in the rather top-down research and development paradigm of the Green Revolution. Variety Release Committees were installed to review the naming and the performance of the varieties that were bred by public breeders. Such systems were developed primarily for the major food crops with special emphasis on the cereals and pulses. The procedures that these committees developed were in most countries included in the seed laws, which initially closely resembled those in the former colonial powers (Bombin, 1980). The first laws made, however, no distinction between different crop groups and the procedures and standards that had been developed for the major food crops were applied to all crops in the law.

In some countries, independent variety testing institutions have been established–often linked to the emerging seed certification agencies–but in most countries, the breeding institutions themselves remained responsible for the execution of the rules: the description of the variety and the collection of performance data.

COMPONENTS OF VARIETY CONTROLS

Variety Registration

Variety registration fixes a variety to a name and includes it in a formal register. Its main task is to identify the variety. Most seed companies have their own register and registration procedures; the name of the variety becomes a brand that has to obtain a position in the market. A national register is primarily meant to support transparency in the market.

The requirements for registration generally depend on the complexity of the market. When few varieties are available, a simple description is sufficient to identify the variety: morphological characteristics such as plant architecture, flower colour, leaf shape are most appropriate to describe a variety, but some agronomic features such as maturity period can be helpful as well.

Identification of varieties of cross-fertilising crops is more difficult because of the genetic diversity within the variety. Depending on the number of varieties in the market, a new variety has to be relatively uniform in order to identify it. The distinguishing characteristics have to be stable after repeated multiplication in order to be able to identify the variety over the years.

In more advanced systems, standard procedures for establishing distinctness, uniformity and stability have been developed for variety registration. The same procedure can also be used to establish whether a variety can be protected (see Ghijsen, this volume).

Next to increasing the transparency in the seed market, an undisputed variety identity is required for the certification of seed lots. The certification system primarily confirms the identity of the variety and the varietal purity.

The data that are necessary may be supplied by the breeder or by an official variety registration agency.

Testing for the Value for Cultivation and Use (VCU)

The decision of farmers concerning which seed they will purchase depends to a large extent on the adaptation to local growing conditions and the product values of the variety. Well-designed variety trials in the region and under representative cultivation practices give the best indication of these values. Coordinated variety trials are the basis of recommended lists of varieties that can be an indispensable source of information for farmers. A coordinated variety testing system requires a central administration and data-analysis, and a decentralised implementation of the trials.

Variety Release Decision

In many countries, varieties are officially released on the basis of the results of variety registration and VCU-testing. This results in a national list of released (or recommend) varieties. Variety release decisions are made by a Variety Release Committee. In order for such committees to work effectively, they have to consist of representatives of all stakeholders (farmers, seed producers, and breeders); in some countries they consist of some independentf 'wise men.'

Formal variety release supposes that it is possible to identify the requirements of all farmers in terms of adaptation to local conditions and cultivation techniques, the value of disease resistances, etc. The committee may review data supplied by the applicant (breeder), by an independent variety testing system, or both.

EFFECTS OF A VARIETY CONTROL SYSTEM

Advantages

A variety control system can be very effective in creating transparency in the seed market. Farmers know what to expect when they purchase seed. Moreover, seed certification systems get an undisputed basis for identifying varieties. Combining registration with performance testing creates a positive incentive for breeding, and gives some level of consumer protection by banning inappropriate varieties from the market.

Countries can furthermore include a plant variety component in integrated pest management initiatives (control of brown plant hopper in Java, Indonesia by using variety groups with different resistance alleles in different growing seasons). Plant variety controls can also be used to secure a premium market for the country's produce, e.g., by allowing long-lint varieties of cotton only in Egypt and Uganda.

One of the most important side effects is, however, that a variety release system based on performance testing creates a wealth of information for farmers. A truly independent and efficient system gives an excellent comparison of the main features of the different varieties in the market when the results are published regularly. Official lists of recommended varieties are a very important tool for farmers to choose their varieties. It is this value that promotes the keen support by both farmers and breeding organisations to support the system in Europe both morally and financially.

The alternative is that farmers rely on the information that the breeders themselves supply. The information from company trials may be as good as the official data, but this is commonly true in competitive seed markets only. Where monopolies or oligopolies exist, there is less pressure to spend as much energy in supplying impartial information on the performance of varieties.

Problems with Registration

Problems with registration are commonly limited to the fact that a registration system fixes the variety to a single description. This means that the variety should not change during its economic life. A breeder is thus not even allowed to make gradual improvements in the variety, which is very well possible in, for example, open pollinated varieties of cross-fertilising crops. At the start of the variety registration system, every cabbage seed supplier in The Netherlands had his own selections from any well-known type of cabbage, and continued to improve these within the general description of the type. The main problem (that these selections were not always clearly distinct) was solved by the introduction of the term 'umbrella-variety.' This concept is, however, very difficult to manage, especially where property rights come in, and breeders have partly for this reason, fully concentrated on hybrids since.

A similar problem arises in developing countries where emerging seed enterprises would like to sell seed of local landraces. Registration of such genetically diverse varieties is technically possible when the diversity can be described and included in the official description. The nature of landraces to respond to changing conditions can, however, not be tolerated. Registration authorities that are currently being set up for the development of Plant Variety Protection have an additional problem with registering landraces, because these do not comply with the stricter rules for protection. It must be clear though that a national variety list (all registered varieties) and a national list of protected varieties are two different things.

Problems with Performance Testing

Problems with the management of variety performance testing relate to Efficiency, Standards, Participation and Transparency (Tripp & Louwaars, 1997).

Lack of *efficiency* includes both inappropriate site selection, and poor trial management. Poor site selection is the result of availability of variety trial stations that are not always situated in the optimum places. These were often established for cash crops, and currently used for all annual crops. Poor trial management results in high residual variances, thus concealing actual differences among varieties, and as a result reducing the number of varieties released or delaying release due to the need for prolonged testing.

To overcome such residual variances, high input levels are used to conceal variation in the trial conditions. Increased soil fertility, careful irrigation and biocide sprays reduce field variation, thus increasing the chance of a higher yielding variety to be recognised. Finally, high input levels give 'beautiful crops' that make a trial credible to visitors. (Trials are commonly weeded very neatly when a high level visitors are expected.) The result is that yield levels in official trials do not represent farmers' conditions at all. Witcombe et al.

(1996) found that average trial yields in the official 1989/1990 sorghum variety trials in India were three times the farmers' average yields (approx. 2800 kg/ha in trials against 900 kg/ha in farmers' fields). It may be doubted whether such trial results have any value to most farmers. The result of the variety release system is that breeders target high input levels, wide adaptation, and easily recognised traits. Private breeders may want to target higher input conditions anyway, because that is where their market is; public breeders who formally focus on remote and resource-poor farmers will not reach their target groups.

Standards relate to the methods of trial layout, data collection and analysis. Most trials contain small plots, often with considerable border effects due to wide paths and other trial design matters, and in any case monoculture plantings, even when farmers plant the crop in combination with others. In many cases yield is the only characteristic that is observed. In the best of cases diseases are screened, but these normally do not go further than marking presence or absence of a particular disease in the field. Quite often, characteristics that are extremely important for farmers are not taken into account, such as aptitude to intercropping, shattering (e.g., soybean), lodging when harvesting is delayed (e.g., maize), cooking time of the produce (e.g., beans). Breeding thus tends to concentrate on yield only without giving credit to the diverse needs of farmers.

Analysis of the trials is often done very objectively, i.e., using standard statistical analysis methods. The main complaint to this analysis is that many trials are pooled in one calculation. The variety having the highest average (!) yield is considered the best. This may not be the best variety in any of the individual testing sites. Standard variety release procedures rarely accept a variety that is specifically adapted to particular conditions, even though they may contain regional recommendations. On-farm variety trials are becoming more and more popular with variety release systems. This very positive development, however, hardly ever contributes to releasing more adapted varieties because such on-farm trials are either completely researcher managed (and thus similar to station trials), or the results cannot be easily analysed statistically. Especially the non-numerical comments by farmers are difficult to include in reports.

Problems related to *participation* relate to the composition of the Variety Release Committees. The committees are commonly dominated by scientific officials (research directors and university professors), in the best of cases supplemented by representatives from the seed production units and in some cases farmers' associations and marketing boards. Even though many committees have identified the stakeholder involvement as a problem, officials often still have the majority vote. In some countries, the high level of officials, and the resulting limited availability, is a serious problem, since it is hardly possible to call meetings. The VRC in Yemen, for example, did not meet for several years in the 1980s and 90s.

Transparency relates to the closed system that formal variety release often is. The policy to support private enterprise often leaves the responsibility for the official trials with the public research institutes. These institutes are not always able to separate their commercialised role as plant breeder and their statutory role as variety testing institution, thus creating a bias against the approval of varieties from the private sector. Attempts to transfer the responsibility to an independent organisation, such as the seed certification service, are commonly contested by the national agricultural research institutes. In addition, parallel (demonstration) trials by the extension service, NGOs or private seed companies are rarely taken into account (Louwaars, 1997).

The International Perspective

Another efficiency problem may arise as the result of the lack of international harmonisation in registration and release procedures and standards. When imported varieties carry a foreign registration report, they commonly do not need to be tested again. One growing season may be sufficient for the certification authority to validate the description under local conditions, which is necessary if the new variety is to be multiplied and its seed to be certified in that country. The description formats in the two countries should then of course be compatible. International harmonisation can reduce the bureaucracy and speed up the introduction of valuable new varieties to farmers. The countries of the European Union have fully harmonised their variety release systems. The registration system is based on UPOV description formats and when a variety is accepted in one country on the basis of performance tests, its marketing will be allowed in all other member states without further ado.

POLICY IMPLICATIONS
OF CONVENTIONAL VARIETY CONTROLS

When the great advantages are taken into account and compared with the identified problems there are two basis responses: either get rid of the system, or allow the system to become efficient.

The general trend is to ease regulations and to allow the market to determine which levels of voluntary controls will develop. This policy is influenced to a large extent by the free market ideology in economic development. Free markets are transparent markets, with educated consumers and with full competition. A voluntary variety control system works quite well in the USA because of the existence of competition in the seed industry, a strong system of universities and experimental stations that perform variety trials, and literate farmers who can make rational choices on the basis of the available information. In developing countries it is much more difficult to generate the appropriate level of self-regulation. Voluntary co-ordinated variety trials will not be funded when one multinational company, which will prefer to spend its money on regional demonstration fields, dominates the market. These demonstrations can be very valuable as a source of information for farmers but they are biased to represent only that company's varieties and hybrids, and to focus on those regions and farming systems that the company can sell seed for.

The option to make the existing compulsory variety testing system more efficient also has some drawbacks. It is relatively easy to improve the formal variety release system in terms of participation. Stakeholders can be given a real voice in national seed boards and variety release committees. Such increased participation is likely to result in a better representation of both on-farm and company trials in the release evaluation. Governments should not fear to give up their majority votes in such bodies. In terms of standards, it is very well possible to change the basis on which performance trials are analysed, i.e., change the current system that a new variety has to be better than the standard into a rule that a new variety should not be worse than the standard. This should go hand in hand with an increased publicity on the performance of all varieties tested. The change of the rule will increase the number of varieties that can be multiplied in the formal seed system (Partap and Sthapit, 1998).

Changing the evaluation standards furthermore promotes a better management of the trials. Nowadays, poor trials with this residual variance results in extremely few varieties show a statistically significant yield increase–when the rule is reversed, such trials will not present a blockade to release.

Finally, governments should regulate only what they can implement. This means that where testing facilities are absent–such as in vegetables in many countries–the variety release system should not be applied. The objective should always be clear: make sure that farmers have access to the best possible varieties. This also means that governments should not want to control varieties in the farmers' seed system, but only those that enter the commercial seed trade. When there is a market for genetically diverse (landrace) varieties, however, the system should not obstruct such initiatives and be able to register such varieties. Also, governments should give responsibilities for the implementation where they are most effectively executed. Variety evaluation that can effectively be handled by the sector itself (e.g., in estate crops), should not be done by an official institution.

RECENT DEVELOPMENTS
TO FURTHER RECONSIDER VARIETY CONTROLS

Intellectual Property Rights

The World Trade Organisation promotes the introduction of intellectual property rights systems for plant varieties (Ghijsen, this volume). Such protection has to be based on a registration procedure that has to determine whether the new application (the variety) is distinct from existing varieties, including all 'varieties of common knowledge' (UPOV-convention). This means that the variety protection authority needs a register of all existing varieties either in a seed bank or in a register of descriptions. These are readily available in countries with compulsory registration systems.

Biosafety

Biosafety regulations in more and more countries require the registration of varieties that contain biotechnologically transferred genes. Such Living Modified Organisms (LMOs) are allowed only when they have passed environmental safety tests, and for food crops that they meet food safety requirements in addition (see Traynor & Komen, this volume).

In addition, a response from consumers in many countries requires that other production systems are guaranteed LMO-free. This requires that such varieties can be readily identified in the market, further supporting the call for a general variety registration system.

Breeding for Diversity/Local Seed Enterprises

Recent strategies for plant breeding for remote and resource-poor farmers include breeding for diversity and participatory plant breeding (Cooper et al. 2001) and the promotion of small seed enterprises (Kugbei et al., 2000). Participatory plant breeding may develop large numbers of varieties that may be selected in small niches only. Such varieties may be quite uniform, but more often genetically heterogeneous. Variety release systems can obstruct such initiatives by disapproving the physical release of varieties that have not gone through the official variety tests.

WAYS FORWARD

Variety registration has an important task in making seed markets more transparent. Performance testing is an extremely valuable source of information for farmers, and has an important task in popularisation of varieties. Variety controls thus support important policies in the seed sector.

It largely depends on the current situation in any particular country, whether seed controls have to be regulated by law and implemented by public institutions, or whether they should be voluntary and primarily implemented by or on behalf of market parties, such as farmers' associations or seed associations.

Variety controls seem to be in a 'catch 22' situation. The perception is that variety controls can only be relaxed when a transparent and competitive seed market exists, while at the same time variety controls are considered a major hindrance in the development of such competitive markets.

In countries that are developing a private seed industry, participation seems to be the main key to breaking the circle. Involvement of all stakeholders is the basis for the development of non-bureaucratic systems that optimally support policy objectives.

Variety registration is an important tool in creating transparent seed markets and in supporting developments in intellectual property rights and genetic modification. Even where registration may be compulsory or voluntary, and procedures may differ there is a very clear advantage in harmonising the technical requirements internationally. Acceptance of foreign registration reports greatly reduces the time needed for farmers to get access to foreign varieties. It seems wise to develop variety registration systems on a regional basis.

Most bottlenecks are identified in the performance testing. Several countries have already reduced the number of crops for which performance tests are compulsory. It seems unwise however to stop coordinated variety trials when these are not required any more for release. Where a legal obligation is absent, it is, however, very difficult to ask breeders for a financial contribution and farmers may not be sufficiently organised to raise the necessary funds. A variety testing system does, however, deliver the necessary performance data that support the farmers' choice and that may balance too wide claims by commercial seed suppliers. Seed associations may be stimulated by government to perform such trials and to make the results publicly available through financial incentives. Regulatory reform in this field may thus not necessarily reduce expenditure.

REFERENCES

Bombin, L., 1980. *Seed legislation*. Rome, FAO.

Cooper, H.D., C. Spillane & T. Hodgekin, 2001. *Broadening the genetic base of crop production*. Wallingford, CABI and London, FAO & IPGRI. 452 p.

Kugbei, S., M. Turner & P. Witthaut (Eds.), 2000. *Finance and management of small-scale seed enterprises*. Aleppo, Syria, ICARDA, 191 p.

Louwaars, N.P., 1997. Regulatory aspects of breeding for field resistance in crops. *Biotechnology & Development Monitor* 33: 6-8.

Partip, T. and B. Sthapit, 1998. Managing agrobiodiversity: Farmers' changing perspectives and institutional responses in the HUH region. Kathucundu International Centre for Integrated Mountain Development/IPGRI. 439 p.

Rutz, H.W., 1990. Seed certification in the Federal Republic of Germany. *Plant Varieties and Seeds* 3: 157-163.

Tripp, R. and N.P. Louwaars, 1997. The conduct and reform of crop variety regulation. In: R. Tripp (Ed.) *New seed and old laws; regulatory reform and the diversification of national seed systems.* London, Intermediate Technology Publications, pp. 88-120.

Witcombe, J.R.W., D.S. Virke and J. Farmington (eds.), 1996. *Choice of seed: Making the most of new varieties for small farmers.* New Delhi: Oxford IBH and London: Intermediate Technology Publications.

The Rules for International Seed Trade

Bernard Le Buanec

SUMMARY. Governments may impose several rules on the seed industry, e.g., on variety registration, seed certification, etc. There is, however,
a whole body of seed regulations that operates outside the direct control
of national governments. These are business rules in international seed
trade that seed producers and merchants conclude among themselves under the auspices of the International Federation for Seeds (FIS). These
rules are aimed at increasing transparency in business, and at reducing
and resolving conflicts. This paper describes the main provisions of this
system. In no circumstance this paper may, however, be used to replace
those rules. The complete rules can be found at <http://www.worldseed.org>.

The international harmonization of these 'Rules for International
Seed Trade' and especially of the definitions used, greatly facilitate the
development of an internationally operating seed industry. Governments
should be aware of the existence of the rules, and of the detailed coverage for seed trade contracts and their conflict resolution through this system. *[Article copies available for a fee from The Haworth Document Delivery
Service: 1-800-HAWORTH. E-mail address: <getinfo@haworthpressinc.com>
Website: <http://www.HaworthPress.com> © 2002 by The Haworth Press, Inc.
All rights reserved.]*

KEYWORDS. Trade rules, international seed trade, business law

THE NEED FOR RULES

Since the very early stage, at the end of the 19th century, the seed trade was
international and still is. Several reasons exist which explain that globalization:

Bernard Le Buanec is affiliated with the International Seed Federation (FIS),
Nyons, Switzerland.

Address correspondence to: Bernard Le Buanec, FIS/ASSINSEL, Chemin du
Reposoir 7, 1260 Nyon, Switzerland.

[Haworth co-indexing entry note]: "The Rules for International Seed Trade." Le Buanec, Bernard.
Co-published simultaneously in *Journal of New Seeds* (Food Products Press, an imprint of The Haworth Press,
Inc.) Vol. 4, No. 1/2, 2002, pp. 143-153; and: *Seed Policy, Legislation and Law: Widening a Narrow Focus*
(ed: Niels P. Louwaars) Food Products Press, an imprint of The Haworth Press, Inc., 2002, pp. 143-153. Single
or multiple copies of this article are available for a fee from The Haworth Document Delivery Service
[1-800-HAWORTH 9:00 a.m. - 5:00 p.m. (EST). E-mail address: getinfo@haworthpressinc.com].

(a) Generally speaking, due to the liberalization of trade starting with the abrogation of the "corn laws" in Great Britain, in 1846, the international trade has increased steadily, this trend being supported and amplified by the development of regional and international trade agreements the most important recent ones being at regional level the European Internal Market, the North American Free Trade Agreement and the Mercosur Agreement and at global level the establishment of the World Trade Organization after the signature of the Marrakech Agreement, in 1994.

During the past 30 years the global seed trade has been multiplied by more than four, to reach presently an annual turnover of almost 4 billion US dollars (see Table 1).

(b) But there are also specific reasons to the seed industry which are noteworthy (Le Buanec, 1996):

- Agricultural production, and thus seed production, is hazardous by nature. In order to ensure regular supply, it is necessary to organize production in different regions so that climatic accidents be avoided. Counter-season growing, which concerns more and more certified seed, allows the acceleration of breeding cycles as well as prevention from seasonal hazards in either of the hemispheres.
- Some agro-climatic zones are particularly favorable to the production of seeds of certain crops: the South-West of France, the North-East of Italy and the Williamette Valley in Oregon for beet seed, Idaho and the high plains of east Africa for beans, the high plains of Central America for flowers, etc. Some countries with climatically favorable zones also have available skilled manpower with an interesting quality/price ratio: South-East Asian countries and Mexico for tomato hybrids, Hungary for maize seed, etc. Finally, for herbage seed, countries such as The Netherlands, New Zealand and Denmark, having a technically and economically very efficient organization of production are to be mentioned.

For these reasons many international production and trade contracts are signed every year. Prior to FIS standardization, parties used to outline their contract specifications based upon local trading usages, leading to misunderstanding, disputes and finally litigations.

TABLE 1. Evolution of the World Seed Trade

	1970	1977	1980	1985	1994	1996	1998
Value in mill. US$	860	1076	1200	1300	2900	3300	3600
1970 basis 100	100	125	140	151	337	383	418

THE DEVELOPMENT OF TRADE
AND ARBITRATION RULES

When in 1924 the first international seed trade congress took place under the auspices of FIS (Leenders, 1967), the agenda was containing three items that are important for this paper: international trading rules, international arbitration and standardization of seed testing methods. The latter was dealt with by the International Seed Testing Association, ISTA, which established in 1931 an international analysis certificate known nowadays as the ISTA International Orange Seed Lot Certificate.

The first edition of the FIS Trade Rules, for herbage seed, was adopted in 1929, only five years after the first congress and the first arbitration rules, in 1930. Since that time, the seed industry has the use of a comprehensive regulatory corpus which has greatly facilitated the development of the international seed trade.

The FIS initiatives were not isolated at that time. The International Chamber of Commerce (ICC) first published a set of international rules for the interpretation of terms in 1936, in order to remedy problems arising form different trading practices in various countries, leading to waste of time and money. Amendments and additions were later made in 1953, 1967, 1976, 1980, 1990 and 2000 (ICC, 2000).

Arbitration has been used extensively in historical times until it was falling into disuse during the 19th century except for very special cases within a family, between neighbours or members of a same organization. However, due to the development of international trade at the end of the 19th century and beginning of the 20th, arbitration knew a spectacular revival and it is now considered as the most favored tool for dispute resolution, particularly in the area of international trade. Two international Conventions were concluded under the auspices of the "Société des Nations" (the League of Nations): a protocol recognizing the validity of the arbitration clause in 1923 and a convention relating to the international enforcement of arbitration awards in 1927. Those two conventions have been considerably improved by a new convention concluded at New York under the auspices of the United Nations in 1958. A European convention was adopted in Geneva also under the auspices of the Unite Nations, dealing with the situation in Continental Europe between liberal and socialist countries made further progress.

THE TRADE RULES

The 2000 edition of the "International Seed Trade Federation Rules and Usages for the trade in Seeds for Sowing Purposes" (FIS, 2000a) was adopted in May 2000. These rules entered into force on January 1st, 2001. They deal mainly with contracts between seed merchants but some paragraphs are also related to contracts with seed growers.

Contracts with Seed Growers

The more detailed information in FIS rules deals with vegetable seed and particularly with growing contracts from stock seed of the contracting buyer. Contracts may be arranged (a) either for an acreage; (b) or for the multiplication of a quantity of stock (basic) seed; (c) or for a quantity of seed fixed in advance. In cases (a) and (b), the total yield must, if corresponding with the norms stipulated in the contract, be delivered by the seed grower and accepted by the buyer. This clause is very important and we know that, if not respected, huge problems may occur as due to unpredictable seed yields, e.g., in grass seed in the USA. If the contract has been concluded for a quantity that had been fixed in advance, the conditions determining the acceptance of any surplus must be established at the time of contracting.

More generally, paragraph 16(c) of the FIS rules relates to multiplication contracts but without details. It is interesting to refer on that issue to the article of Chopra and Chopra in Fenwick Kelly and George (1998). Whilst presenting the experience gained in the development of "seed villages" during the Green Revolution in India, the various possibilities are interesting and also used in other countries. Four options for procurement contracts are presented.

Fixed Price

The procurement price of processed seed is negotiated in advance and fixed on a weight basis. The seed is procured only if it meets the prescribed field and seed standards as certified by satisfactory grow-out tests (when appropriate) and a recognized seed testing laboratory. This option is usually preferred for "costly" seeds, such as hybrids and vegetable seeds.

Premium Over Grain Price

This option is well adapted to high volume, low price seed such as rice, wheat, cowpea, open pollinated maize, sorghum or sunflower. The procurement contract establishes a premium over the average wholesale price in a recognized grain market for the 7 or 15 days previous to the delivery of the seed. The premium may vary from 6 to 15% according to the crop and the terms of delivery and payment.

Profit Sharing Through Pool

The growers take an advance equivalent to 50% of the grain price and keep their seed in the company pool. The company deducts the costs involved in processing, storage and marketing, plus a previously agreed profit margin, usually 10% for the quantity sold. The balance is paid to the growers after deducting advance payment in proportion to their quantity in the pool. The unsold stocks are either carried over, sold as grain, or returned to the growers. This option is well suited to small farmer-owned seed enterprises and to developing countries.

Premium Over Negotiated Price

To encourage contract growers to produce seed of higher quality than the minimum standard, a higher price is paid for better quality seed, taking into account among other characteristics germination and genetic purity. (In fact that fourth option is not an option in itself but an additional one which can be added to the three first ones.)

Contracts Between Seed Merchants

The FIS Trade Rules consider the following issues in detail.

General Provisions

They contain definitions of terms of time and communication such as "hour," "day," "working day" and "telecommunication." Most importantly they also indicate that, when the code letters "FIS" are embodied in a contract pertaining to seed, the FIS rules apply in full, except if it is specified otherwise in writing. These code letters are also valid as arbitration clause (cf. *infra*).

Conclusion of a Contract

A description of the conditions must be given in the contract, including the following: the date, the quantity, the species and the variety, the description of the quality, the price per unit, the type of packaging, the delivery period, the payment conditions and particular provision when appropriate. The rules for making the contract binding are also indicated. It is important to indicate if the contract is concluded "subject to an import or export authorization," "subject to crop" and "guaranteed to pass" or "subject to passing." All these conditions are explained in the rules.

Conditions of a Contract

This chapter gives detailed information on how to describe the quantity, with a conversion table between kilogram and pound, and the quality with definition of terms like "about," "approximately," "minimum," "maximum" and the possible tolerances. It also deals with the packaging, the shipping terms based on the "INCOTERMS" (cf. *supra*), the documents which have to be presented and the insurance.

Execution of the Contract

The shipment and the payment are the two important parts of the execution of a contract.

The shipment may be made either at "buyer's option" or scheduled before a "fixed date" or within a "predetermined period." In case of fixed date or within a predetermined period, the seller has to inform the buyer of the intent to ship in advance, the time frame being subject to the distance. If shipment is done at buyer's option, the buyer must give notice in advance, time limits vary according to the rules such as "immediate shipment," "prompt shipment," etc. The delays of shipment and the default of shipping instructions and the default of shipment are also considered in detail in the rules.

Unless other specifications are provided for in the contract, the payment has to be made net against document (cf. *supra*) at first presentation. An annual interest rate is provided for if the payment is not settled in due time.

Quality Checks and Analyses

The declaration of the quality of the seed can be made in one of the following ways:

- by the furnishing of an official seed testing report
- by the furnishing of a seed testing report other than an official one, issued by a governmental or a private laboratory
- by a simple declaration.

The choice must be specified in the contract. The FIS rules also indicate how the buyer and the seller have to behave in these respective cases, in particular in case of contestation by one of the parties.

Disputes

Any complaint regarding a difference of weight, packaging, appearance, moisture content, specific purity and the specification of the seed (including grading and coating) must be made at the first discovery of the defect and within a maximum of 12 working days after the arrival of the seed at destination.

Any complaint regarding the germination must be made at the first discovery of the inferiority and within a maximum of 60 days after the arrival of the seeds at destination.

Any claim concerning defaults of trueness to variety or varietal purity shall be made within normal delays of sowing and of control in the region of the buyer and, at the latest, within a maximum delay of one year after receipt of the seeds by the buyer.

If the dispute cannot be settled amicably or by mediation or conciliation, it shall be settled by binding arbitration according to the FIS Procedure Rules for Dispute Settlement, with the exclusion of ordinary judicial procedure unless otherwise expressly agreed to in writing. Application for arbitration shall be made within 30 days from the date of breaking off the friendly negotiations or from the date of the termination of the mediation or conciliation process.

DISPUTE SETTLEMENT

The 2000 edition of the International Seed Trade Federation (FIS) Procedure Rules for Dispute settlement for the Trade in Seeds for Sowing Purposes and for Management of Intellectual Property (FIS, 2000b) provide for alternative dispute settlement procedures: mediation, conciliation, arbitration.

A mediation is a negotiation carried out with the assistance of a neutral third party–the mediator–who does not have the authority to give an award or to impose a decision on disputing parties. The mediator acts as a facilitator, just helping them to reach an agreement.

A conciliation is a process in which the neutral third party–the conciliator–not only attempts to motivate parties toward a final settlement, but also could be asked to give the parties a non-binding opinion. This opinion remains confidential.

An arbitration is a process in which each party presents its case to an arbitration tribunal for a final and often binding decision (binding in case of FIS Arbitration).

Why Arbitration?

Independently of the economic structure of the various countries there is not, at the moment, any general agreement on a competent tribunal to deal with disputes in international trade. This lack of agreement on which national laws should apply in case of disputes. Moreover, those national laws, that were primarily made for domestic trade, are often not very appropriate to solve international disputes. Arbitration is, therefore, used to remedy that defect in international order. An arbitration clause is now included in almost all contracts in international trade. That clause firstly means that the parties agree on arbitration to settle their disputes in case of disagreement, secondly it specifies the competent arbitration tribunal and the relevant arbitration rules and thirdly it stipulates the "international trade law" which should apply. In case of seed trade most of the contracts include an "FIS" arbitration clause referring to FIS trade rules, arbitration rules and arbitration chambers.

Another good reason to use arbitration in international trade disputes is that the normal courts have difficulties to take decisions on very technical matters. They have to ask the opinion of experts. It is, therefore, more efficient to address directly those experts in the frame of a well established procedure.

Is Arbitration Legal and Widely Accepted?

The two main questions regarding the effectiveness of arbitration are: Is an arbitration clause valid and are the arbitration awards enforceable?

Generally the parties comply on their own free will with the arbitration clause and the arbitration award for the sake of their commercial reputation. However, in order to avoid any difficulty when exceptionally the parties don't comply on their free will, numerous international treaties and conventions have greatly assisted in establishing a predictable legal environment for the enforcement of the arbitration clause and the award. The most important treaty (cf. *supra*) in international arbitration is the United Nations Convention on the Recognition and Enforcement of Foreign Arbitral Awards of 1958, known as the New York Convention. One hundred and fifteen countries are signatories to the New York Convention. It addresses the recognition by domestic courts of arbitration clauses, the mandatory referral by such courts to arbitration pursuant to the clause, the judicial enforcement and recognition of arbitration awards (exequatur) and the grounds for refusal of such recognition and enforcement. This regime has become so successful that in most countries it is now far easier to enforce a foreign arbitration award than a foreign court judgement. Information on this system can be found at <http://www.internationaladr.com/te.htm>.

The FIS Arbitration Procedure Rules

This section describes general rules for arbitration. An expedited procedure exists in FIS Rules, which may apply when the financial claim has a contractual ground, and does not exceed the amount of CHF 100,000.

The Arbitration Clause

As indicated in *supra* any contract established according to FIS rules and with the inclusion of the code letters "FIS" provides for mandatory binding arbitration, unless otherwise expressly agreed upon in writing. These code letters are equivalent to an arbitration clause. If no reference to FIS is made in the contract and if the parties agree however on FIS arbitration, an arbitration clause such as the following should be included: "All disputes arising in connection with the present contract shall be finally settled under the arbitration procedure rules of the International Seed Trade Federation."

To make the arbitration possible, the arbitration clause is essential.

Organization of the Arbitration

In conformity with FIS Articles of Association, member associations must organize facilities for conducting arbitration either directly or through an existing Arbitration Chamber. Each member association must establish a list of arbitrators in its country. Only arbitrators appearing on that list can be appointed to Arbitration Committees. The arbitrators have to respect the "Code of Ethics for Arbitrators" published by FIS. In particular they must not conduct themselves as lawyers of the parties.

Application for Arbitration

Application for arbitration must be sent to the Arbitration Chamber of the seller's country or, if in that country no Arbitration Chamber has been organized yet, to the Secretary General of FIS who must select an arbitration chamber in a third country.

The application must contain, among other information, the precise summary of the points in dispute and the object of the application.

Application for arbitration must be made within the time limit specified in the FIS Rules and Usages for the Trade in Seeds for Sowing Purposes (cf. *supra*).

Nomination of Arbitrators

Each arbitration is dealt with by 3 arbitrators, nominated from the list of arbitrators established by the Arbitration Chamber. Each party nominates one arbitrator and the third arbitrator is nominated by the Arbitration Chamber.

An arbitrator can be challenged by a party under the law of the country in which the arbitration is to be held. A reasoned written request must be sent to the competent Arbitration Chamber. If the challenge is accepted by the Arbitration Chamber, the party who nominated the challenged arbitrator shall nominate a new arbitrator.

The Arbitration Procedure

The parties are invited to attend a hearing, organized by the Arbitration Chamber. The parties may attend personally or may be represented by members of their own or another member association of FIS or by a duly accredited proxy. At the request of arbitrators the parties must supply all the details and information regarding the case.

Then, after consideration, the arbitrators draw up the arbitration award which must contain, among other information:

- the names of the arbitrators,
- a description of the matter in dispute,
- a statement of the facts, the decision and the ground for the decision,
- the amount of the cost and who is to pay it, and
- the binding signature of the arbitrators.

The award must comply with the provisions of the Conventions of New York and Geneva (cf. *supra*).

If the arbitration award is not subject of appeal, it is binding on the parties and enforceable. "Exequatur" may be requested to a national court if needed.

Appeal

If the parties disagree with the award, each of them may lodge an appeal to the Secretary General of FIS who will charge a new Arbitration Chamber in a third country which is not the country of any of the parties. The appeal is accepted provided that the party having lost the case furnishes a security to guarantee that the award of the first instance will be implemented if confirmed. The possibility of appeal gives the party insurance of fairness in dealing with their case. The obligation to furnish a guarantee for payment prevents the appeal to be a stalling tactic.

The appeal award is final and binding and the obligations falling upon the parties have to be fulfilled within 30 days that follow the receipt of the award.

If an award is not implemented, the party that has won may demand enforcement according to the rules of the New York Convention. In addition the Secretary General of FIS notifies to all member associations the party that has not fulfilled its obligations.

CONCLUSIONS

Seed is one of the most regulated products by comprehensive seed laws at national level. The OECD Seed Schemes from 1957 onwards have greatly facilitated and still facilitate the international movement of seed in helping the seed industry to fulfil their regulations.

However, in order to help the seed industry in implementing the rather complex international seed business, it has been necessary to develop a corpus of "commercial laws" including trade rules and arbitration rules. FIS started its activity in that area as soon as 1924 and was successful in implementing the first set of rules in 1930. Thus FIS was one of the pioneers in the establishment of such an organization which, is not unique in the world, but still belongs to the few really world-wide specialized trade ruling and arbitration institution. FIS is proud of this achievement which certainly is one of the reasons of the development of the international seed trade.

REFERENCES

Fenwick-Kelly, A. & R.A.T. George, 1998. *Encyclopedia of seed production of world crops*. Chichester, New York: John Wiley, 603 p.

FIS, 2000a. *International Seed Trade Federation (FIS) Rules and Usages for the Trade in Seed for Sowing Purposes*, Nyon, FIS.

FIS, 2000b. *International Seed Trade Federation (FIS) Procedures Rules for Dispute Settlement for the Trade in Seeds for Sowing Purposes and for Management of Intellectual Property*, 2000. Nyon, FIS.

ICC, 2000. *INCOTERMS 2000*, ICC Publishing S.A., Paris.

Le Buanec, B., 1996. Globalization of the Seed Industry: Current Situation and Evolution. Seed Science and Technology, 24, 409-417.

Leenders, H., 1967. The Function of the International Seed Trade Federation (FIS) in the International Seed Trade, Proc. Int. Seed Test. Ass. Vol. 32.

OECD, 2000. *OECD Seed Schemes for the Varietal Certification or the Control of Seed Moving in the International Trade*. Paris, OECD.

WTO, 1995. *Trends and Statistics*, International Trade.

WTO, 1999. *FOCUS No. 43*, Nov. 1999.

Progresses in the Turkish Seed Industry

Nazimi Açikgöz
Canan Abay
Nevin Açikgöz

SUMMARY. The Turkish seed industry is in a rapid change and privatization is going on as it has been planned. Recently, 20,400-hectares of state farms have been announced for joint ventures on seed production. At the moment, almost 92% of wheat and 83% of barley seed production is done within the public sector. The state farms run by TIGEM (General Directorate of Agricultural Enterprise) are the main seed suppliers of the public sector.

Ecological advantages of Turkey for seed production have been evaluated positively and hybrid seed production for export, especially for corn and sunflower, is gradually increasing. On the other hand, vegetable seed imports are also rising due to increased demands for export of greenhouse vegetable products. In this case, the government's policy is to promote seed companies to produce their seeds in Turkey itself.

Field trials of genetically modified corn, cotton and potato varieties have begun in Turkey two years ago, and the biosafety regulations are under development.

Seed policy specialists could select the Turkish seed market as a laboratory, especially from the standpoint of privatization and transformation of local seed companies.

Although Turkey is a member of ISTA and OECD, its membership to

Nazimi Açikgöz is Professor and Canan Abay is Assistant Professor, Seed Technology Center, Ege University, Kampüsü, 35040 Bornova-Izmir, Turkey.
Nevin Açikgöz is Associate Professor, Agean Agricultural Research Institute, P.O. Box 9, Menemen-Izmir, Turkey.
Address correspondence to: Nazimi Açikgöz at the above address.

[Haworth co-indexing entry note]: "Progresses in the Turkish Seed Industry." Açikgöz, Nazimi, Canan Abay, and Nevin Açikgöz. Co-published simultaneously in *Journal of New Seeds* (Food Products Press, an imprint of The Haworth Press, Inc.) Vol. 4, No. 1/2, 2002, pp. 155-163; and: *Seed Policy, Legislation and Law: Widening a Narrow Focus* (ed: Niels P. Louwaars) Food Products Press, an imprint of The Haworth Press, Inc., 2002, pp. 155-163. Single or multiple copies of this article are available for a fee from The Haworth Document Delivery Service [1-800-HAWORTH 9:00 a.m. - 5:00 p.m. (EST). E-mail address: getinfo@ haworthpressinc.com].

UPOV is yet underway. The harmonization of "Variety Development, Registration and Release" and "Seed Quality Control and Certification" regulations with the European Union has also been started. The Seed Technology Center at Ege University, Izmir has been established as a specialized institution. *[Article copies available for a fee from The Haworth Document Delivery Service: 1-800-HAWORTH. E-mail address: <getinfo@ haworthpressinc.com> Website: <http://www.HaworthPress.com> © 2002 by The Haworth Press, Inc. All rights reserved.]*

KEYWORDS. Seed sector development, seed legislation, seed research, institutions, Turkey

INTRODUCTION

At the beginning of the 1920s, Turkey established state farms to transfer the new agricultural technologies to the farmers. The plant breeding institutes, agricultural research stations and agricultural faculties and universities were established from 1923 onwards. The first result of formal plant breeding, the barley variety 'Tokak,' was registered in 1937, which is still being used as the main source for the improvement of two-row barley varieties. The results of agricultural research have contributed considerably to the agricultural development of the country. In cereals, the yields per unit area have doubled over the last 70 years.

The state farms under the frame of TIGEM (General Directorate of Agricultural Enterprise) are the main seed suppliers to the public for self-pollinating crops, but are increasingly coming under the ambit of the privatization policies of the government. Privatization of the state farms can, however, cause a disruption of the seed supply because private seed companies usually prefer cross-pollinated crops particularly hybrids and not self-pollinated crops, due to absence of breeder's rights.

Turkey has 77 million ha area, of which 24 and 3 million ha are occupied by agricultural and horticultural crops, respectively. Production statistics of cereals comparing with the other agricultural crops in Turkey, are given in Table 1.

The relative importance of the breeding activities, leading to the use of modern varieties, and of the seed industry varies greatly per crop (Table 2), which shows the relative difficulty of the seed sector to supply seed of self-pollinating crops.

The Ministry of Agriculture and Rural Affairs (MARA) has formed an Agricultural Council in 1997 and initiated a series of reforms in seed regulation and administration, which has led to the establishment of the Seed Advisory Committee composed of all seed related institutions in Turkey.

TABLE 1. Production information of field crops in Turkey.

	Area Sown (1000 ha)	Percentage in Cereal Sown Area	Percentage in Total Area	Production (1000 Ton)	Average Yield (kg/ha)
Wheat	9,350	67.5	52.0	17,000	1,818
Barley	3,350	24.0	19.0	6,500	1,940
Rice	62	0.5	0.4	162	2,612
Maize	600	4.0	3.0	2,000	3,166
Total	13,362	96.0	74.4	--	-
Indust. Crops	2,300	Total = 77 million ha			
Legumes	1,600	Forest = 21 million ha			
Fodder Crops	600	Grassland = 13.5 million ha			
Follows	5,200	Urban = 15.7 million ha			
		Field crops = 23.5 million ha			
Total Field Crops	23,500	Horticulture = 3.2 million ha			

TABLE 2. Area sown, coverage of modern varieties, seed demand and use of certified seed of some major crops.

Crops	Cropped Area (1000 ha)	Area of Modern Varieties (%)	Seed Demand (1000 T)	Certified Seed (%)
Wheat	9,350	75	370	25
Barley	3,350	25	140	10
Rice	62	13	8	4
Maize	600	22	12	24
Sunflower	500	100	3	100

VARIETY DEVELOPMENT, REGISTRATION AND RELEASE

Almost all 58 public institutions in agriculture, and 10 private seed companies, are involved in plant breeding and development of varieties for the 17 major agro-ecological zones in Turkey. Beside the registered varieties, varieties licensed for production exist as well. The Registration Committee meetings are arranged for plant groups in spring time, every year and the list of registered varieties is published in the official gazette. In the year 2000, registration committees approved 69 varieties, with the largest number in wheat. There are more than 100 registered wheat varieties and not all of them are in seed propagation program in Turkey. Although the number of registered varieties seems to be high for some plants, variety improvement is a continuous work and requires very strong support from the government. Due to non-existence of intellectual property rights on plant varieties yet, a protection system is in development and membership of the International Union for the protection of New Varieties of Plants (UPOV) is on the agenda of the parliament. Passing of the law will be required to increase the role of the private sector in variety development and fulfilling the requirements of the TRIPs Agreement which is necessary for membership of the World Trade Organisation.

To evaluate the large number of wheat varieties on the national list, it is pertinent to realize the ecological diversity of the country plus the diversity in farming systems. Seventeen macro-ecologies for cereals can be distinguished in Turkey. The number of wheat cultivars for 2000 seed propagation is 79, whereas the number of barley is only 24 (Table 3). In this case each wheat variety has an average coverage of 9,350,000/79 = 118,000 ha, whereas this figure is 140,000 ha for barley. For one maize cultivar only 5,000 ha and for one rice cultivar 3,000 ha is the average potential market for seed production.

Seed Production and Distribution

The Ministry of Agriculture and Rural Affairs (MARA) coordinates the seed production in Turkey and both public and private sectors are involved. Agricultural research institutes and enterprises produce breeder and basic seed, whereas the public and private seed companies and nucleus farms (contract farmers for seed production) produce certified and controlled seed. Certified seed is distributed through public institutions, agricultural credit cooperatives and agencies. In the recent privatization efforts, a public institution, Agricultural Equipment Agency (Zirai Donatim Kurumu) had to cease its seed distribution operations.

Seed Quality Control and Certification

Ankara University had carried out seed certification until the Seed Control and Certification Institute was established in 1959. A year later, the Regional Field Trial and Release Institute was established (including five regional seed

TABLE 3. Number of registered varieties and number of the varieties subject to seed propagation (NVSSP) in Turkey.

Crops	Total # of Listed Varieties	# of Varieties in NVSSP	Cropped Area ('000 ha)	Average Area for Each Variety ('000 ha)
Wheat	123	79	9,350	118
Barley	39	24	3,350	140
Rice	26	18	62	3.5
Maize	123	121	600	5

certification laboratories in Samsun, Istanbul, Antalya, Izmir and Mersin). In 1963, the variety release, seed quality control and certification law was promulgated (Article 308) and Turkey also became a member of ISTA. Three years later, sugar beet, cereals, maize, sunflower, soybean, and fodder crops certification were included into the OECD Seed Scheme.

In 1986, the Seed Registration and Certification Center (SRCC) was established following the amalgamation of Regional Field Trial and Release Institute and Seed Release, Control and Certification Institute, with the mandate for overall variety registration, seed quality control and certification of agricultural crops in Turkey. In 1998, the SRCC activities covered 72 ha of field controls and 146,000-ton seed and almost 4 million samplings for certification. During the seed multiplication pre- and post-control tests were carried out on 33% of exported and 10% of locally used seed.

Changes in Agricultural Production

The cropping pattern in Turkey is changing rapidly in the light of new export opportunities and changing consumption patterns:

- new export facilities are leading to major increases of durum wheat production for macaroni products, and vegetable production for export, e.g., to the new republics of the former Soviet Union,
- increasing living standards and changing consumption preferences are increasing the consumption of processed food.

The seed sector has to consider such issues carefully, because they have to find the suitable cultivars for the given ecology and produce the needed seeds.

Nowadays, Turkey is in a transition period for each sector including seed industry because of the globalization and integration with European Union.

Share of Private Sector in Seed Production and Distribution

Until 1980s, almost all seed production and distribution used to be in the public sector except sugar beet. Thereafter, different types of private seed companies (multinational, national or joint ventures) have started to take part in seed business. Table 4 provides the share of private sector for the years 1995 and 1999 for some crops. It is obvious that with realization of plant breeders' right, the public sector can withdraw itself from the seed market much further like in western Europe and North America.

Turkey has a specific ecological advantage in seed production. For example, maize seed can be dried to below 14% moisture content in the field, which facilitates healthy and cheaper seed production. All major maize seed companies produce seeds in Turkey and export to their European partners.

TABLE 4. Share of private sectors in Turkish seed market for the years 1995 and 1999 for some crops.

Crops	1995 (%)	1999 (%)
Wheat	3	8
Barley	4	17
Sunflower	99	100
Potato	99	100
Cotton	1	14
Fodder Cr.	11	41

In 2001 there were about 80 national and international seed companies in Turkey. Almost half of them are now member of Turkish Seed Industry Association (TURK-TED) and those members are covering almost 90% of the Turkish seed market. The Association participates in all seed related activities of Ministry of Agriculture, represents its members in the State Planning Organization, the Union of Chambers of Agriculture, the Seed Consultation Committee, the Seed Registration Committee, etc. Seed related bureaucratic procedures like variety release and registrations, certification and issuing import-export permits are within the domain of four General Directorates of Agricultural Ministry.

Seed Research and Training

Turkey has established a well-equipped Seed Technology Center (STC) at Ege University, Izmir capable of conducting national and international seed courses. A course on seed pathology was organized for the private sector recently. There will be a national training course on comparison of international quarantine regulations.

The Center thus has an important impact on the further development of both public and private functions in the seed industry.

The Center has a national consulting committee with participation from the Ministry of Agriculture, private seed sector and universities. In the recent meeting, the Center was asked to prepare national strategies and policies for the seed business in Turkey.

Agricultural Biotechnology and the Seed Sector

Field trails for genetically modified crops were started in 1998 for maize, cotton and potato. Turkey is not the center of origin of these crops and environmental hazards are, therefore, limited. The transgenic crops are potentially useful for Turkish agriculture as they reduce production costs–especially the use of insecticides. In order to avoid major export problems, it is important, however, that EU regulations on transgenic crops be followed strictly.

The contributions of transgenic crops can best be explained by maize and cotton, which are grown on 550,000 ha and 700,000 ha, respectively. Access to European corn borer resistance variety may provide opportunities for double cropping of maize after barley or wheat on an estimated area of 100,000 ha with a net benefit of approximately $1000 per ha. Cotton production is labor intensive and needs heavy insecticide application, a minimum of 5 to 10 sprays in western Turkey and 15 sprays in Adana region. The amount and cost of insecticide and environment consequences are of major concern. If cotton growers in western Turkey get access to insect resistant cotton varieties available on world market, it would be possible to meet the national raw material need for the textile industry in a sustainable manner.

Agricultural Policies

According to an analysis by the IMF (International Monetary Fund), Turkey has to take important steps in changing agricultural support policies, to rehabilitate agricultural unions and privatize state tobacco and alcohol monopolies. Also the Turkish Republic Agricultural Bank (TC Ziraat Bankas), which is a basic supporter of farmers, will not be "the bank of farmers" any more. New reform processes based on a "direct income support" system have been launched, supported by the World Bank, but the first experiences seem unsatisfactory.

Urgent Needs of New Seed Policies and Regulations

Many years of support policies included also the seed industry. A list of supported crops with their amounts are presented in Table 5. This list was changing every year. Alternatively, seed companies started to support their customers with credit schemes for up to one year. Contract-based production seems to be the most promising solution.

According to the seed law 308 all seed control measures are the responsibility of the Ministry of Agriculture. Within that ministry, however, there are four different General Directorates, and coordination and harmonization of seed control are in need of rehabilitation. The European example where the seed certification and seed tests are being run by an autonomous body like GNIS in France or NAK in Holland may be followed. This would not oppose the policy of privatization of services. Bureaucratic tendencies can be well illustrated with the Quarantine Services that prescribe a zero tolerance for Mosaic Virus contamination in imported alfalfa, where such quality is not available in the world market.

TABLE 5. List of supported crops based on seed and support amount.

Crops	Tl/kg[1]	Crops	Tl/kg
Rice	100,000	Fodder beet	100,000
Soybean	100,000	Chickpea	100,000
Cotton	100,000	Bean	150,000
Potato	50,000	Peas	200,000
Vetch	100,000	Tomato (Hybrid)	300,000,000
Clover	500,000	Cucumber (Hybrid)	10,000,000

[1] These figures are based on 1,000,000 Tl = approx. 1 US$.

The informal sector is likely to shrink with time but the marketing of non-certified seed can not be easily controlled with existing regulation. Reform new measures are urgently needed. Even though the seed law 308 gives legal options for restricting such marketing, the above-mentioned lack of coordination is a major obstacle for its implementation. In some cases, illegal seed trade rises up to 80%, such as with cotton seed in the year 2000.

The implementation of "plant breeding rights" and membership of UPOV may benefit the Turkish seed business. At the moment, many international seed companies and also many promising varieties are not entering into the country. The uncertainty in ownership and trade is the main handicap. It may seem that these companies are losing, but the Turkish farmers are definitely losing more.

An urgent legal change is required which enables seed producers to start small-scale seed firms in remote and isolated areas. A special regulation should facilitate such firms to cover their own quality control needs.

In Western counties, public agricultural research institutes sell their varieties and not the seed. This is unfortunately not true in Turkey. This is due to various reasons and it is not very conducive for the development of a healthy seed industry. Therefore, legal changes seem to be indispensable.

As discussed earlier, the private seed sector in Turkey is new, and they are of three types: (i) 100% foreign owned, (ii) 100% Turkish, and (iii) a hybrid ownership. Suitable ecological condition of Turkey for seed production need to be harnessed for benefit. The privatization of the state farms offers new opportunities for seed company development and could contribute to the seed export goals.

CONCLUSION

The Turkish seed industry is making rapid progress and successful transformation. Privatization is continuing as planned by the government. Seed export is growing with contributions from the private sector created by the free market economy. The multinational seed companies will bring the latest technology provided UPOV membership is finalized. At present, importing agricultural biotechnology products is accepted, regulations for field-testing approved, and regulations for production and use are under development.

Worldwide, the farmers are getting benefit from agricultural biotechnology. In 2000, within five years period, the area under transgenic crops reached 43 million ha worldwide. Turkey may be the first country in the region to make use of transgenic crops. Exchanging basic knowledge and information among the policy makers, scientists and seed specialists is essential during discussions for preparation of regulations on biotechnology.

The Seed Technology Center in Izmir is the only specialized seed institution in Turkey, with over 50 specialists conducting seed technology training and research. The Center had organized national/international congresses, symposia and meetings and is planning a series of seed-centric activities for Rural Affairs and the Turkish Seed Industry Association in the future.

So far globalization of the seed sector is advancing rapidly. But what would be the future of local seed companies? Such issues need to be studied in more detail in Turkey as well as in other WANA countries. Therefore, a comprehensive research project is required to develop seed strategies for WANA countries.

REFERENCES

Açikgöz N., B. Eser, H. Saygili and F. Sarvan. Anforderung an die Sorten für den Getreideanbau in der Türkei. Paper presented in symposium for "42. Jahrestagung der Geselschaft für Pflanzenbauwissenschaftten" 10-12/9/1998 Freising-Weihenstephan Deutschland.

Açikgöz N. Possible contribution of transgenic cultivars to the Turkish agriculture. Paper presented in "1999 World Seed Conference" (6-8 September, 1999) Cambridge, UK.

Acikgoz N. & N. Acikgoz. Transgenik Çesitler, Islahc Haklar Ve Terminator Teknolojisi.Anadolu 9(1) s: 95-194, 1999. (Transgenic varieties, intellectual property rights and terminator technologies.)

Acikgoz N. Possible contribition of transgenic cultivars to the Turkish agriculture. Paper presented in "1999 World Seed Conference" Cambridge, UK.

Acikgoz N. The National Seed Sector in Turkey. Seed Info. No. 18, 2000, pp. 10-12, (WANA-ICARDA).

Seed Industry Development
and Seed Legislation in Uganda

Fred Muhhuku

SUMMARY. Agriculture is the predominant economic activity in Uganda and the government's vision is to develop a profitable, competitive, sustainable and dynamic agricultural and agro-industrial sector. To achieve this requires that the majority smallholder farmers access yield-enhancing technologies. Seed is regarded as a crucial input since it is the basic means of technology transfer to farmers.

Policy development for research and seed production and control is put in a historical context, showing the need to regularly update policies, institutions and laws in order to meet the requirements of the changing conditions. Smallholder focus through stimulating participatory plant breeding and stimulating private investment in the formal seed sector need to be included. Furthermore, the composition of policy making and implementing bodies changed, and regulations were adapted with a focus on regional harmonisation.

Seed policies, laws and regulations will need to be reviewed again to accommodate new trends in the industry such as GMOs, genetic conservation, and biosafety. A suitable regulatory framework is crucial in strengthening the seed industry. And a successful seed industry will be instrumental in the government's drive to modernise agriculture in Uganda. *[Article copies available for a fee from The Haworth Document Delivery Service: 1-800-HAWORTH. E-mail address: <getinfo@haworthpressinc. com> Website: <http://www.HaworthPress.com> © 2002 by The Haworth Press, Inc. All rights reserved.]*

Fred Muhhuku, at the time of preparing the paper, was affiliated with Uganda Seed Projects headoffice, Kawanda, Uganda.

Address correspondence to: Fred Muhhuku, ADC/IDEA Project, 18 Prince Charles Drive, P.O. Box 7856, Kampala, Uganda.

[Haworth co-indexing entry note]: "Seed Industry Development and Seed Legislation in Uganda." Muhhuku, Fred. Co-published simultaneously in *Journal of New Seeds* (Food Products Press, an imprint of The Haworth Press, Inc.) Vol. 4, No. 1/2, 2002, pp. 165-176; and: *Seed Policy, Legislation and Law: Widening a Narrow Focus* (ed: Niels P. Louwaars) Food Products Press, an imprint of The Haworth Press, Inc., 2002, pp. 165-176. Single or multiple copies of this article are available for a fee from The Haworth Document Delivery Service [1-800-HAWORTH 9:00 a.m. - 5:00 p.m. (EST). E-mail address: getinfo@haworthpressinc. com].

KEYWORDS. Seed policy, institutional change, technology transfer, biosafety, genetic resources, informal seed sector, privatisation, harmonisation

BACKGROUND

Agriculture is the predominant economic activity in Uganda, involving 2.5 million farm families. It contributes about 40% of GDP and 90% of export earnings. It also accounts for more than 80% of employment, and 60% of total government revenues.

The Government of Uganda agricultural policy objectives are:

- Increased agricultural productivity to ensure food security and self-sufficiency in raw materials for agro-processing and surplus for export;
- Increasing household incomes and reducing peoples' poverty; and
- Diversifying the country's exports through promotion of non-traditional export crops.

The national vision is to have a profitable, competitive, sustainable and dynamic agricultural and agro-industrial sector by the year 2017 (Government of Uganda, 2000). To achieve this, government recognises that growth in the sector will depend on introducing technological change and ensuring that smallholder farmers adopt yield-enhancing technologies. Seed is regarded as a crucial input since it is the basic means of technology transfer to farmers.

In the colonial times crop production was divided into two categories, namely cash crops and food crops. Most of the cash crops were grown on large estates (except coffee and cotton) and all had their own elaborate systems for seed production and distribution. The private sector also played a key role in both research and seed multiplication. On the other hand, smallholders using their own landraces grew food crops on a subsistence level.

In 1968, the government started a seed scheme in the Ministry of Agriculture to provide improved seeds for the traditional food crops, so as to uplift the productivity of the majority smallholder farmers. A network of agricultural research stations, which had been earlier established to spearhead public research in crop improvement and development, was already generating improved varieties. Between 1971 and 1986 political instability and economic mismanagement totally ruined the cash crop economy, while food crop production informally became a significant source of income for most farmers.

DEVELOPMENT OF THE SEED INDUSTRY

Agricultural Research

For all the food crops, variety development is the responsibility of the public agricultural research system within the Ministry of Agriculture. Started over 100 years ago, a number of excellent varieties had been developed before the politico-economic ruin referred to above began to take its toll. Thus revival came through the creation of the National Agricultural Research Organisation (NARO) in 1992.

The Strategic Plan for Agricultural Research prepared in 1991 defined the organisational framework for NARO's establishment, and provided guidance on its legal status, governance mechanisms, and research resources management systems. NARO brought together a fragmented set of institutes within a single structure, as a semi-autonomous body under the Ministry of Agriculture, Animal Industry and Fisheries (MAAIF). It was established by *The National Agricultural Research Organisation Statute*, 1992, as a body corporate with perpetual succession and a common seal.

The objects of NARO are to undertake, promote, and streamline research in agriculture, livestock, fisheries and forestry. The Organisation comprises of a governing board, a secretariat, specified research institutes, and associate institutes which may be approved from time to time.

Since its establishment, NARO has generated a number of crop technologies including new, high-yielding and disease resistant varieties of cassava, beans, potatoes and maize that are widely used by farmers. Also in use are new varieties of bananas, horticultural crops, pigeon peas, finger millet, sorghum and groundnuts. NARO works closely with International Agricultural Research Centres, like CIAT, CIMMYT, IITA. Most of the seed for these crops, however, is supplied by the informal sector, with the formal sector accounting for less than 10%. For instance average annual seed sales of maize is 1000 MT which is approximately 12% of national seed requirement, while beans is 800 MT which is about 6%. In the case of vegetatively propagated crops, all planting materials are supplied by the informal sector.

NARO is continuing to refine its strategies and is seeking effective participation of stakeholders in research planning, technology assessment and dissemination in order to make technologies more relevant to producers and consumers. One of the key strategies, therefore, is to increase the efficiency and effectiveness of technology development and dissemination through the greater participation of producers in all the stages of the research and dissemination process. The private sector is also being encouraged to participate in the funding and provision of agricultural research and related services. At the same time, NARO is decentralising research services, with devolution of responsibilities to the Agricultural Research and Development Centres (ARDCs) located in the various agro-ecological zones.

Seed Production and Distribution

Due to the general economic problems the seed scheme was never able to fully address the farmer needs for improved seeds. Revival came in 1983 when a grant obtained from the EU was used to rehabilitate the basic seed infrastructure including farms, processing equipment, laboratories, as well as manpower development. Uganda Seed Project was established with the responsibility to multiply and distribute improved seeds of the major cereals, legumes and oil-seed crops. The project was also responsible for seed quality control.

Following this phase, government obtained a loan from the African Development Bank in 1993 to rationalise the various efforts in the seed sector and hopefully stimulate the commercialisation of the industry. Lack of a suitable regulatory framework was hindering development of the industry despite gonvernment's policy of economic liberalisation (see Box 1). The specific objectives of the ADB funded phase were:

- To further increase the production and distribution of high quality seeds of improved varieties of the stated crops,
- To separate seed quality control from the production and marketing,
- To commercialise the activities of seed production, processing and marketing, and eventually privatise them, and
- To put in place a seed law and make it operational.

THE SEED REGULATORY FRAMEWORK

The Seed Statute

The Agricultural Seeds and Plant Statute was enacted in 1994. The statute provides for "the promotion, regulation and control of plant breeding and variety release, multiplication, conditioning, marketing, importing, and quality assurance of seeds and other planting materials and for other matters connected therewith." To achieve this, various organs were provided for, namely: The National Seed Industry Authority, The National Seed Certification Services, and The Variety Release Committee.

The National Seed Industry Authority (NSIA)

This is a multidisciplinary board with the overall responsibility to oversee the development of the seed industry. It is expected to advise government on

---

BOX 1. Grappling with a liberalised seed industry in a legal vacuum

By 1993, the overall Government policy was one of "economic liberalisation," and the seed industry was no exception. The problem, however, was that there were no legal instruments to guide investment in the sub-sector. As a result, a number of trans-national companies, interested in entering the Ugandan market got cold feet and withdrew after initial attempts, e.g., Pioneer Hibred and Cargill. Kenya Seed Company (KSCo) on its part resorted to operating informally by selling seed across the border into Eastern Uganda where agro-ecological conditions are similar to those in Kenya.

At the same time, a number of NGOs and International Relief Agencies wanted sizeable quantities of seed for their operations in Uganda and within the region, which Uganda Seed Project was not able to supply. These Agencies included UNHCR, ICRC, World Vision, CARE International, Lutheran World Federation, Catholic Relief Services, etc. As a result, a number of local trading companies swung into the seed business, buying ordinary produce from the market, processing it and selling it as (certified) seed! With the liberal policy and no clear laws and regulations, these companies did brisk business, and the NGOs did not mind provided that "the seeds germinate." Some of the companies involved in this trade have since evolved into formal seed companies and are registered with the government certifying agency, e.g., Harvest Farm Seeds (a ubsidiary of Commodity Export International), and Farm Inputs Care Centre (a subsidiary of AFRO-KAI Ltd.).

Another method the NGOs used to obtain seed was to organise informal seed production using farmer groups. This was done in West Nile region by CARE International, in Hoima-Kibaale region by a church based NGO supported by the Belgian Government and in Rakai district with funding from the Irish Government. The Belgian NGO in Hoima has since been converted into a commercial company and registered with the certifying agency as Nalweyo Seed Company (NASECO), while in Rakai the farmer groups are now producing certified bean seeds on contract to various commercial seed companies. Farmer groups in different parts of the country are also still the sole producers of planting materials for vegetatively propagated crops like cassava and potatoes. The revised seed statute will seek to formalise such groups.

---

the various aspects of seed legislation affecting multiplication, distribution and marketing of seeds in Uganda. The specific functions of NSIA are:

- Formulating and advising the government on national seed policy (at the moment there is no written policy and this is expected to be the starting point of the authority's work);
- Establishing a system of implementing seed policies through technical committees;
- Reviewing the national seed supply and advising government on the administration of the seed industry; and

- Coordinating and monitoring the public and private seed sectors in order to achieve the national seed programme objectives.

The Authority is composed of senior officers from the Departments of Crop Production, Forestry, Trade and Marketing, and Animal Resources. In addition, there are representatives of farmers, seed growers, seed traders, and the cotton industry. The Minister for Agriculture appoints two other members. The Head of Crop Production chairs it, while the Director of Seeds is the secretary. It is obvious that there is over-representation of government officials on the Authority at the expense of the private sector, which has often raised concern among the private players in the industry.

The National Seed Certification Services (NSCS)

The overall mandate of the NSCS is to design, establish and enforce certification standards, methods, and procedures. The specific responsibilities are:

- Receiving, adjusting, maintaining and enforcing seed standards established through research;
- Advising NSIA of any modifications to seed standards and providing them with information on any technical aspects affecting seed quality;
- Providing training to persons responsible for the implementation of the statue; and
- Registration and licensing of all seed producers, conditioners and dealers.

To fulfil these responsibilities the NSCS would handle the following routine functions: variety testing, field crop inspection, laboratory testing and issue of official seed certificates, inspection of seed processing and storage facilities, labelling and sealing of processed seed, inspection of marketing outlets, issue of import/export permits, issue of licenses. The NSCS also provides facilities as the secretariat of NSIA.

The Variety Release Committee (VRC)

This committee is composed of the Head of Agricultural Research (Director General, NARO) as the chairman, and the members include: all Directors of the Research Institutes, the General Manager Produce Marketing Board, a representative of Seed Traders, and one from the Ministry of Trade and Industry. The functions of the VRC are:

- To review and maintain the national variety list and to approve new varieties of seeds;

- To review the history and performance records of selected varieties of seeds;
- To determine the contribution of varieties of seeds for agricultural development;
- To approve variety release and entry of seeds into the seed multiplication programme;
- To make recommendations of obsolete varieties of seeds;
- To determine varieties of seeds to be released, rejected, referred or outclassed;
- To establish standards of varieties of seeds eligible for seed certification; and
- To give advice to plant breeding organisations on market and farmers requirements.

Other Elements of the Statute

The statute also gives detailed guidelines on the following aspects:

- Plant breeding and the registration of breeders;
- Seed multiplication and the licensing of seed producers;
- Seed conditioning and the licensing of conditioners;
- Seed trade, import and export;
- Seed sampling, testing, labelling and sealing of bags;
- Offences and penalties of offenders; and
- Some exemptions.

The statute briefly touches on phytosanitary standards and practices, as well as the granting of plant breeders' rights. Six schedules and a set of regulations give details of the standards, procedures, and rules to follow in implementing the statute.

Difficulties in the Implementation of the Seed Statute

There was delay in the implementation of the statute, causing sections of it to be overtaken by certain developments both within Uganda and from outside. It soon became apparent that implementing it in its present form would not be possible because of the following developments:

- Restructuring and reorganisation of the public service led to the merging of some departments and ministries. Certain offices were abolished and titles/designations of officers changed. This affected the composition of the organs of the statute.

- The liberalisation of the entire economy opened up the seed industry leading to entry of several companies into the seed business. The ability of the government certification agency to cope with this was suddenly called to question, and yet there was no provision in the statute for delegation of certain responsibilities to the seed merchants themselves, or to any other body. While some organisations/companies exploited weaknesses in the non-operational law to make quick money by selling substandard seed as certified seed (see Box 1), the Uganda Seed Project often used sections of the statute to frustrate private companies it regarded as competitors (see Box 2).
- The revival of existing regional trading blocks, and the creation of new ones, in Africa, e.g., East African Community, COMESA, SADC, created opportunities for seed companies to operate in larger, more profitable markets. It thus became necessary to harmonise relevant laws and regulations in countries forming those trade blocks in order to enhance seed trade. The Association for Strengthening Agricultural Research in East and Central Africa (ASARECA) has already gone along way in such an exercise for the EAC countries (Kenya, Uganda and Tanzania). Inevitably certain sections of the Ugandan seed law were affected.
- The free market economics in Uganda, coupled with the opportunities presented by regional trade attracted transnational seed companies into Uganda. However, these companies were concerned about protection of their varieties in the absence of suitable PVP legislation in Uganda. The government, keen to attract private investment in agriculture, moved quickly to address this by appointing a Task Force to draft a suitable law. *The Plant Varieties Act* has now been drafted, which necessitates deleting certain sections in the seed statute.

Revision of the Seed Statute

In view of the above developments, and in order to breathe new life into the statute, it has been agreed to review and update it. The exercise has been facilitated by DANIDA through the Agricultural Sector Programme Support (ASPS) for provision of regulatory services, which is a key function of the Ministry of Agriculture, Animal Industry and Fisheries. The review will aim to strengthen the regulatory function of the government as seed production is largely taken over by the private sector. The following areas described below have been reviewed.

Seed Laws and Regulations

The aim is to ensure that laws and regulations governing seed production and certification do not prohibit local seed production, foreign investment in the industry, and regional trade. Following are some of the changes:

- The name of the statute has been changed to *The Seeds and Plant Statute* to enable it encompass all types of seeds.
- The criteria regarding the registration and licensing of seed merchants, conditioners, dealers, and growers were considered and clarified. For instance, registration of seed growers is the direct responsibility of seed merchants who then pass on the information to the NSCS. This greatly reduces the workload of the certifying agency.
- The section dealing with phytosanitary standards and practices was deleted as these are well covered by the Plant Protection Act. Similarly, appropriate reference has been made to the draft Plant Variety Protection Act.
- A section dealing with appeals has been introduced to cater for anyone not happy with decisions of the certifying agency. The Seeds and Plant Tribunal has been established as a final stage of appeal before an aggrieved party resorts to the High Court.
- The informal seed sector has been acknowledged and allowance has been made for the NSCS to train and offer technical guidance to bodies involved in it.
- In line with all the changes made, appropriate modifications have been made in the Regulations, the Schedules and the various Forms. For instance, in Schedule 3, the Classes of Seed were changed in conformity with the harmonisation of seed regulations in East Africa.

BOX 2. Some examples of misuse of the seed law to frustrate private seed companies

The law requires that a new variety should be tested in Uganda for at least 3 seasons before it can be officially released and registered in Uganda. This section of the law has sometimes been used to deny Kenya Seed company an import permit even when it is common knowledge that their varieties have always been grown in eastern Uganda where no suitable locally bred ones exist.

Apart from testing for the 3 seasons, the law also requires data on multi-location, on-farm and DUS tests but does not specify whether these tests should be done concurrently or in separate seasons. National breeders do the first two while a unit within Uganda Seed Project does DUS. Often some of this data is not available, delaying the release of a potential variety by several years and frustrating the efforts of commercial seed companies.

The VRC is convened by Uganda Seed Project and sometimes takes several years without sitting on the pretext that there are not enough applications to warrant a meeting. The law does not specify the number of applications needed to convene a VRC meeting.

Sustainability of the Seed Certification Services

How the seed quality control and certification services can be managed, both in the short- and long-term, on a self-sustaining basis was considered. A section was introduced dealing with the charging of fees for all the services carried out by the certifying agency. Other sources of funding for the NSCS were specified for the avoidance of any doubt, e.g., government budgetary allocations, grants and donations.

Allowance has also been made for the delegation of certain responsibilities to other organisations and private seed merchants, as well as accreditation procedures. This will reduce the burden on the certifying agency and improve efficiency in the delivery of services.

Operational Procedures for Variety Release

The activities of the Variety Release Committee, variety testing, registration and release were reviewed and changes made in accordance with the harmonisation process among the EAC countries. The changes have made the procedures less cumbersome and time consuming which has been discouraging the private sector. For instance, the number of seasons of testing a variety before official release has been reduced from three seasons to only one.

A provision has been made for the formation of a Variety Performance Technical Committee to conduct national variety performance trials in conjunction with the relevant breeder. This committee will advise the National Variety Release Committee so that the actual process of releasing and registering a new variety becomes a mere formality.

Composition and Functions of the Organs

The name of the National Seed Industry Authority (NSIA) has been changed to National Seed Board (NSB) as the title "Authority" was conveying inaccurate connotations in contemporary Uganda. The composition of the Board and its functions were reviewed and strengthened in line with the changes in the seed industry. For instance there is more representation from the private sector. At the same time, the correct titles of the government representatives have been used according to the restructured Civil Service.

The composition of the National Variety Release Committee has been changed to give greater representation of the private sector (including private breeders). Removing the Director General NARO from the chair has further reduced dominance of this committee by the Public Researchers. Instead, the Minister will now appoint the chairman from among all prominent Ugandan agricultural scientists.

THE WAY FORWARD

The NARO Statute is being well implemented, with appropriate strategies being taken to adopt modern trends like, participatory variety development and dissemination, decentralisation of services, and greater role of the private sector in research. The draft PVP law has been completed and should be passed into law soon. Membership of UPOV is being considered and a Biosafety Committee has been set up in the Uganda National Council for Science and Technology (UNCST). All this is aimed at stimulating research and development of new varieties, as well as attracting transnational seed companies to invest in the Ugandan seed industry.

The government has taken the correct move in separating seed quality control from production, and withdrawing from the latter, which gives the private sector a greater role in the commercial aspects of the seed business. The revised seed statute is an important instrument in the development of the seed industry and should be operationalised without further delay. The changes made will remove the remaining bottlenecks in fully liberalising the seed industry. Though the overall Uganda Government policies are very progressive compared to most sub-Saharan countries, the problem has been the failure to implement them. However, in spite of these shortcomings, there are already excellent varieties developed by NARO, while several private seed companies (both local and international) are becoming active in Uganda, for instance:

- Farm Care Inputs Centre, Harvest Farm Seeds, and NASECO are local private companies multiplying and distributing locally bred varieties, while General & Allied is marketing assorted vegetable seeds under its own label "AFRISEEDS";
- Kenya Seed Company and East African Seed Company have been locally incorporated in Uganda and market their own varieties; and
- Trans-national companies like Seed Co. International, Pannar, and Monsanto are testing and marketing their varieties in the country.

The Seed Policies, Laws and Regulations will need to be reviewed regularly to accommodate new trends in the industry. For instance, the National Council for Science and Technology addresses issues concerning GMOs, genetic conservation, and biosafety. A suitable regulatory framework is crucial in strengthening the seed industry. And a successful seed industry will be instrumental in the government's drive to modernise agriculture in Uganda.

REFERENCES

Agrar Consulting GMBH. Einbeck, 1992. *Seed Industry Rationalisation Project: Seed Marketing and Privatisation Study.* Consultancy Report, Ministry of Agriculture, Entebbe, Uganda.

Government of Uganda, 2000. *Ministry of Agriculture, Animal Industry and Fisheries Policy Statement 1999/2000.*

Government of Uganda, 2000. *Plan for the Modernisation of Agriculture: Eradicating Poverty in Uganda.* Government of Uganda. August 2000.

Government of Uganda, 1999. *Plan for Modernisation of Agriculture: Eradicating Poverty in Uganda* (Draft Document, November 1999).

Government of Uganda, 1994. *The Agriculture Seeds and Plant Statute, 1994.*

Government of Uganda, 1992. *The National Agricultural Research Organisation Statute, 1992.*

Gwarazimba, V. E., 1999. *Uganda Seed Sub-Sector Review* [Report, July 1999].

Kabeere, F., 2000. *A Study on Harmonisation of Seed Policies, Laws and Regulations in Eastern Africa: The Ugandan Seed Industry Component.* Ministry of Agriculture, Entebbe.

Lavery, P., 2000. *Revision of the Seeds and Plant Statute and Regulations. Entebbe, Ministry of Agriculture, Animal Industry and Fisheries.*

National Agricultural Research Organisation, 2000. *A Strategy for 2000-2010: Facing the Research Challenges for the Modernisation of Agriculture.* NARO, Entebbe, May 2000.

Uganda Seed Project. *Various Reports of the Uganda Seed Project/Seed Industry Rationalisation Project.*

Seed Regulatory Frameworks
in a Small Farmer Environment:
The Case of Bangladesh

Md. Nazmul Huda
Hans W. J. Smolders

SUMMARY. Seed markets dominated by small resource-poor farmers are usually more difficult to penetrate because of low purchasing power, farmer's immobility and information barriers. Farmer's subsequent reliance on own or locally traded seed does render farming systems sustainable but generally prevents significant yield increase because of a slow influx of new varieties and lack of quality guarantee. Seed regulatory frameworks in such small-farmer environments should, therefore, strongly focus on removing barriers for variety introduction and exchange and lay less emphasis on the control functions.

This article describes the development of the seed regulatory framework in Bangladesh and gives a detailed account of the current framework introduced in 1993, followed by an assessment of the impact 8 years after implementation.

The new seed regulatory framework has had a significant impact on the level of private sector investment in the country. Yet, investments are found mainly in the downstream seed supply, initiated by the introduction of a voluntary system of seed certification. Further observations show that farmers have better access to new varieties in crops that are ex-

Md. Nazmul Huda is Seed Industry Consultant, Building Number 2, Flat Number B-4, 2 Lake Circus, Kalabagan, Dhaka-1205, Bangladesh.

Hans W. J. Smolders is affiliated with the Bangladesh-German Seed Development Project, Dhaka, Bangladesh.

Address correspondence to: Hans W. J. Smolders, Vlietstroom 23, 3891 EM Zeewolde, The Netherlands.

[Haworth co-indexing entry note]: "Seed Regulatory Frameworks in a Small Farmer Environment: The Case of Bangladesh." Huda, Md. Nazmul, and Hans W. J. Smolders. Co-published simultaneously in *Journal of New Seeds* (Food Products Press, an imprint of The Haworth Press, Inc.) Vol. 4, No. 1/2, 2002, pp. 177-193; and: *Seed Policy, Legislation and Law: Widening a Narrow Focus* (ed: Niels P. Louwaars) Food Products Press, an imprint of The Haworth Press, Inc., 2002, pp. 177-193. Single or multiple copies of this article are available for a fee from The Haworth Document Delivery Service [1-800-HAWORTH 9:00 a.m. - 5:00 p.m. (EST). E-mail address: getinfo@haworthpressinc.com].

empted from government control. The frequency of variety introductions in the remaining five controlled crops has not significantly increased. As these crops represent more than 90% of the country's total crop seed value, it is concluded that the impact of the new framework on variety introduction and hence on agricultural output has been relatively small. In view of the present situation, a review of the seed regulatory framework and its monitoring functions is urgently needed. *[Article copies available for a fee from The Haworth Document Delivery Service: 1-800-HAWORTH. E-mail address: <getinfo@haworthpressinc.com> Website: <http://www. HaworthPress.com> © 2002 by The Haworth Press, Inc. All rights reserved.]*

KEYWORDS. Seed policy, seed institutions, seed regulatory framework, small farmers, farming systems, variety control, Bangladesh

INTRODUCTION

Seed regulatory frameworks generally aim to provide the farmer community with a continuous supply of new varieties in an effort to enhance crop yield and quality and increase the farmer's income. Live seed is the means of this supply.

Bangladesh is a typical small farmer country, which causes seed to be a vital element of rural life. More than 50% of its 125 million people are engaged in agriculture, and of this about 40% consist of landless sharecroppers. With little options available for employment in other sectors, with a growing population, landownership gradually decreases in size. At about 0.4 ha per farmer household this is already one of the lowest in Asia (World Bank & BCAS, 1999).

At 600,000-ton annual seed requirement, the size of the seed market in Bangladesh is large. Of this volume less than 6% is supplied by the formal sector. Small farmers often include high yielding rice and other crop varieties in their cropping system, despite their inability to purchase certified seed. This shows that they do not resist new developments, rather they continuously demand new varieties in all major and minor crops. In order to deliver on this demand, national seed systems require not only a sound regulatory framework, but also one whose implementation is strongly pursued and supported by government.

FRAMEWORK CONDITIONS
IN SMALL FARMER COMMUNITIES

Farmers in poor countries often have less access to innovations than those in more developed countries. Lower purchasing power, immobility, local power structures and illiteracy hamper their access to new varieties and the necessary information thereof. Consequently, small farmers use their own seed or locally traded unlabelled seed as a safe and easy source of seed, which enables them to sustain their farming systems and to source new varieties. While this informal seed supply guarantees availability of seed in the farmer's proximity, it does not normally compensate for major weaknesses in the system such as the lack of quality guarantee and the slow influx of new varieties. In countries troubled by high rainfall and floods like Bangladesh, the seed quality in this seed system fluctuates tremendously year by year as revealed by Huda (1992). Yet, small farmers have an excellent testing ground for new varieties. They habitually grow a range of varieties simultaneously to sustain a minimum level of production and to suit production to various output purposes.

It is, therefore, not difficult to see why informal seed systems compose a formidable competition to any formal sector trademark supplier. This is particularly the case in the cereal crops rice and wheat, although recent studies in Indonesia have revealed that similar competition also exist in crops, like soybean, that are less likely to be retained at farmer level (Udin et al., 1995). Since labelled seed usually is more expensive, trademark suppliers need to provide a significant higher product value to small farmer communities in terms of yield benefit, quality and service.

Without a strong commitment to provide farmers these benefits, farmers may not achieve the progress foreseen. This fact should be anticipated by facilitating the smooth flow of seed and varieties and allocate sufficient funds to research and other seed institutions in support of the sector.

Like many other developing countries, Bangladesh has moved away from strict variety control and since a couple of years enjoys a new seed regulatory framework (Bangladesh Government, 1993). Although the new policies actively support private sector development, the liberalisation has only been partial as variety introduction of five major crops continues to be under the scrutiny of the government. Sources say this was a compromise decision as considerable fear existed to enter unknown territory and that the liberalisation would threaten the existence of the national research institutes (Gisselquist & Srivastava, 1997, see Box 1).

Has the move been sufficient to offset major changes in the national seed system or is it necessary to anticipate on a more radical restructuring? This article aims to give a brief picture of the historic development of the Bangladesh seed regulatory framework and to assess the impact of the new seed policy, now nearly 8 years after introduction.

Historic Development of the Framework

Prior to 1974 there was virtually no seed regulatory framework in Bangladesh. Activities like varietal development of rice started in 1908 with the es-

BOX 1. The case of a local seed company

Mr. Sarwar Jahan Azad is the managing director of a local seed company in Sherpur, 150 km north of the capital Dhaka. His company is among an array of new seed companies emerging in Bangladesh, who are benefiting from the new seed regulatory framework which recently came into being. Less than a decade ago, government regulations on seed production and trade was strongly discouraging private sector investment and gave virtual market monopoly powers to the Government Seed Corporation BADC. Ironically, it is the BADC nowadays that helps the company of Mr. Azad to establish his business by providing most of the hardware for seed conditioning (seed cleaning, treating and storage) against service charges. Key to his and many other business is the government's policy change from mandatory to voluntary certification, which has sparked many seed initiatives in Bangladesh. Public to private sector cooperations, like that between BADC and Mr. Azad's company was anticipated in the Seed Policy 1993, gradually seems to take form as similar service modules are emerging throughout the country.

---

tablishment of a 400 ha research farm at Dhaka; later on a second research farm attached to the Agricultural College in 1938 was founded. Conscious attempts to produce and supply seed were only taken in 1954 through the establishment of 20 Seed Multiplication Farms. In 1961, all input supplies including seed were separated from the Department of Agriculture by creating the Bangladesh (formerly East-Pakistan) Agricultural Development Corporation (BADC). The quality of the seed supplied by the corporation was, however, often poor, which attracted serious criticism from the users.

The World Bank undertook a study on the seed situation in Bangladesh in 1969 (World Bank, 1973). The study identified many gaps and proposed a project with the name Cereal Seed Project. However, implementation of the project was delayed due to the liberation war in 1971 and post-war reconstruction priorities. The Cereal Seed Project activities finally started in 1974, envisaging the creation of a regulatory framework.

From 1974 onwards many important events took place while many points were added to the development of the country's seed regulatory framework (Table 1).

The National Seed Board (NSB) and the Seed Certification Agency (SCA) were established in 1974 as the highest seed policy making body and the authorised institute for seed certification, respectively. Simultaneously, seed production through contract seed growers and seed processing through seed processing centres were started. The seed ordinance was promulgated in 1977 and the seed rules in 1980. With this regulatory framework in place, formal seed production, processing and marketing started. Priority in development was given to wheat and rice seeds; the system gradually expanded to jute seeds and seed potato.

TABLE 1. Chronology of events in the development of the seed regulatory framework

Year	Event
1974	• Establishment of Seed Certification Agency.
	• Creation of National Seed Board.
1975	• Establishment of Contract Growers' Zone.
	• Establishment of Seed Processing Centre.
1977	• Promulgation of 'The Seed Ordinance 1977.'
1980	• Promulgation of 'The Seed Rules 1980.'
	• Seed Certification Manual approved by NSB.
1981-82	• Seed certification of cereal seed started and expanded also to jute seed.
1990	• Private sector and NGO participation was considered necessary to increase the volume of quality seed and to develop seed industry.
1990-93	• Seed Policy was drafted, finalised, approved in NSB and cabinet.
1993 March	• The Seed Policy was notified in the gazette.
1997	• Seed Ordinance was amended according to the Seed Policy.
1998	• New Seed Rules called 'The Seed Rule 1998' promulgated cancelling the Seed Rule 1980.

FORMATION OF THE SEED POLICY 1993

During the late 1980s when the input supply of the fertiliser and irrigation sectors, until then handled by BADC, were privatised, the question of privatising seed supply emerged as a point of discussion. Since management of seed activities was different than those of fertilisers and irrigation equipment, different strategies had to be pursued. This prompted for the formulation of a Seed Policy.

The situation prevailing during the late eighties regarding seed activities, which lead to the formation of the Seed Policy, can be summarised as follows:

1. The quantity of seed supplied by the BADC was inadequate to meet the national requirement.
2. Involvement of the private sector was limited to low-volume high-profit seeds, mainly vegetable seed.
3. The supply of seed through BADC was highly subsidised by the government.
4. A growing awareness among farmers of the benefit of using quality seed.

5. Many seed traders expressing their willingness to undertake seed business.

The development partners working at that time with seed, like GTZ, World Bank, The Netherlands, and FAO, suggested measures to restructure the national seed system. Germany (GTZ), supporting the seed sector since 1976, suggested for the creation of a separate seed corporation with financial and administrative autonomy. When this did not materialise GTZ's position lead to the discontinuation of the German seed programme in 1989. GTZ renewed its support to the seed sector only in 1997. The World Bank suggested the commercialisation of the BADC seed operations, which would allow for more transparency of the seed costs and gradual withdrawal of subsidies (World Bank, 1988). The Bank also proposed that BADC would provide direct support to the private sector. Both elements were included in the Seed Policy.

For the preparation of the Seed Policy, the MOA formed a committee, which considered the present (then) situation and the long-term outlook of the sector development, reviewing some of the Seed Policy documents of the neighbouring countries. The drafted Seed Policy was first approved by the National Seed Board and then by the Cabinet of the Minister (for excerpts see Table 2).

SEED POLICY IMPLEMENTATION

Although the draft was available in 1990, formal adoption of the Seed Policy took almost three years. One of the reasons behind this delay was the non-availability of the Seed Wing as the implementing agency under the Ministry of Agriculture. The Seed Wing was only established in August 1992. The other reason for the delay may have been the non-cooperative approach of the public sector, particularly the research institutes and the seed producing agencies (BADC), who regarded the upcoming private sector as incapable and as competitor (Gisselquist & Srivastava, 1997). A strong commitment of the Ministry towards seed sector re-structuring finally made it possible to pursue the formalisation of the seed policy.

After notification in the government gazette in January 1993, the Seed Wing took up the activities related to the implementation of the Seed Policy by holding a seminar with around 300 key participants to discuss the implications in detail. Although the participants were divided in their view, the seminar produced a document with 18 recommendations, but it fell short of a full plan of action.

After adoption of the Seed Policy it became necessary to modify the Seed Ordinance 1977 and the Seed Rules 1980. A taskforce consisting of members

TABLE 2. Excerpts of the Seed Policy 1993

Seed Policy 1993 Excerpts

General Objective

1   The overall purpose of the Seed Policy is to make the best quality seed of improved varieties of crops conveniently and efficiently available to farmers with a view to increasing crop production, farmers productivity, per capita farm income and export earnings.

Regulations

2   Originally breeders' seed was only made available to the BADC; with the adoption of seed policy, samples of breeders' seeds are also made available to the private seed entrepreneurs.

3   Truthfully labelling of seed: In the seed policy, certification is made voluntary and a class "truthfully labelled seed" is created where seed producers are allowed to attach their own 'truthful' label on the supplied seed bags or packets.

4   Certification of breeder's seed and foundation seed: Certification of breeder's seed and foundation seed is made compulsory and SCA is named the only agency to certify these two classes of seed of all crops bred by the public sector.

5   Controlled crops: The National Seed Board (NSB) decides which kinds and varieties of crops are to be (de)notified (in other words controlled). Five crops remain notified: rice, wheat, jute, potato and sugarcane. To market any new variety of these crops, the variety must pass through a testing procedure finally to be approved by the NSB.

6   Other crops: Any variety or hybrid of the crops other than the five notified crops can be marketed by registered dealers after giving a declaration about its useability whereafter it will get a number from the NSB.

7   Registration of seed dealers: All intending seed entrepreneurs must register with the secretariat of the NSB as seed dealers in order to obtain the authorisation to undertake seed business.

8   Control of quality of the truthfully labelled seed: Field officers of the SCA obtain authorisation to collect samples of the marketed seeds for testing.

9   Import of seed: Except for appropriate plant quarantine safeguards restrictions on importation of seed will be eliminated. Approved varieties of notified crops may be imported for commercial sale. However, registered seed dealers will be permitted to import small quantities of seeds of notified crops for adaptability testing.

Role of Institutions in Seed Sector

10   Role of BADC; The BADC will be run on a commercial basis as far as possible and will be re-oriented to promote the development of the private sector seed industry through advise, training, rendering its facilities for use to private sector.

11   Role of Seed Wing/Ministry of Agriculture; Favourable policies, incentive and supports will be provided to promote private sector participation in seed industry.

from public sector research organisations, BADC, the Seed Wing (MoA) and the president of the Seed Merchants Society (private sector) was created to undertake the preparation of the modification. The draft modification was processed and finally adopted by the parliament in the Seed (amendment) Act 1997 and the Seed Rules 1998 by the Ministry. Later on, a plant quarantine rule was drafted which is in the process of formalisation. Also a plant variety protection law is currently in the process of drafting.

IMPACT OF THE SEED POLICY

After eight years of implementation, it is worthwhile to take a look at the impact of the Seed Policy. Policies need time to take effect and cannot be seen in isolation from other regulatory processes in the agricultural and industrial sector, with which it forms a cohesive course of action.

The Changed Role of Seed Certification

Starting from 1974, seed certification and quality control has been vested on the Seed Certification Agency. For a long time, however, certification could not be implemented as there was, beyond BADC, no organised seed multiplication and marketing system in place, and BADC's internal quality control system was technically more capable to handle the issue of field inspection of contract seed growers and seed testing. Later an agreement was reached with BADC to allow formal inspection of the BADC declared seed grower plots and testing of declared seed lots kept in the seed processing centres. These procedures were followed until the Seed Policy 1993 was published.

In the Seed Policy 1993, certification of seed was made voluntary, whereas certification of breeders' seed and foundation seed remained compulsory. Another class of seed named 'truthfully labelled seed' was introduced to allow seed producers and companies to test and certify the seed themselves.

By freeing the system from compulsory pre-market controls, it was considered necessary to introduce a market quality control system at wholesale and retailer level. The new Seed Rules 1998 authorised the SCA to conduct this new task. Lately the country was divided into 32 zones and responsibilities have been assigned to designated SCA officers to collect seed samples from wholesalers and retailers, inspect the bags and undertake testing in the laboratory. Clear cut procedures to conduct market inspections and imply stop-sale authority, however, are still under consideration, which require the endorsement of the NSB.

Up until now the private seed sector therefore enjoyed a considerable freedom to explore and activate its development, which has been a much needed incentive to spur seed sector investment. Initial attempts of the SCA to impose market control recently have met resistance with private producers and retailers, who look at such attempt as policing, and fear their investments being threatened.

The capacity of the SCA also is barely enough to cover the mandatory duties of certification, thus leaving little manpower for nationwide market control and advisory duties. Although occasional reports of bad seed quality and inappropriate use of labels are reported, the misconduct has been proportionally small. As such it is felt that policing the sector in this early stage of private sector development will have a negative effect on the sector's progress. Apparently, the conversion of the SCA from a controlling body to a service oriented institution is difficult.

Impact on Variety Release and Registration

The new Seed Policy strongly pushed forward liberalisation of the crop improvement sector by excluding many crops from notification. Effectively, any non-notified crop variety can be imported freely just by applying for a registration number, which is granted without testing. Only five crops namely rice, wheat, jute, potato, and sugarcane, which are regarded sensitive to the national agriculture and food security, remained notified. For the marketing of notified crops, all varieties including hybrids are subject to prescribed formalities for release and registration.

In order to get the varieties of notified crops approved and released, the originating breeder or company has to submit an application to the NSB, after which the variety enters into a series of multi-locational trials. This variety release procedure takes a minimum of two years, but experience is that it frequently lasts longer. The variety is finally released by a notification in the official gazette. This procedure was in place before 1993 and is, with slight modification, still practiced.

A technical committee and field inspection team, consisting primarily of public sector research and extension officials is charged with the supervision and reporting for release. Procedures for variety release are cumbersome and needs the active lobbying even by public sector breeders. There is no independent body for variety testing, both for Value for Cultivation and Use (VCU) and for Distinctness, Uniformity, Stability (DUS) testing.

Although the Seed Policy is a significant step forward, assessment of the impact on variety release indicate that little progress is made. During the past 8 years, no private company has put forward any crop variety for notification. Hybrid rice, which is a recent introduction to Bangladesh, is the only exception to this fact. Because of its high potential to national output, the government significantly accelerated the release procedure. Four hybrid varieties imported from India and China were released in 1998 on the condition that the private companies should start production of hybrid rice within 3 years time. Although there are now more than 30 hybrid varieties under testing, of late, only 2 varieties have been released.

The intensity of the release of varieties has not or barely increased since the Seed Policy was published (see Figure 1). In certain crops like wheat and potato, the number of varieties released during the last 5 years remained the same or decreased.

Among the rice varieties released during the last 10 years, only 2 varieties have been well accepted by farmers. Some varieties that seemed to do well, like BR-31 and BR-32, the demand suddenly defected due to a specific character not liked by the farmers.

In wheat, the variety Kanchan nearly entirely dominates the national crop production. Similarly, varieties of Cardinal and Diamond, both originating from the Netherlands, dominate the potato crop production. With few new varieties released in these crops, the likelihood of increased crop output in Bangladesh in the immediate future is limited. Rather, it is likely that because of a break down of disease and pest tolerance or resistance, yields may decline.

FIGURE 1. Intensity of Variety Introduction Since 1960

Intensity of Variety Introduction

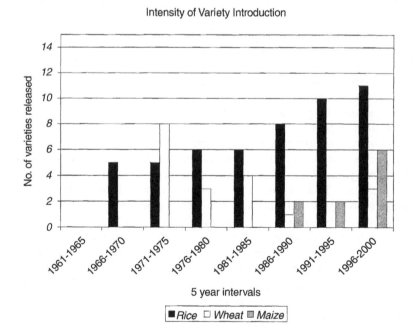

5 year intervals

■ Rice □ Wheat ▨ Maize

Exemplary of the weakness of the variety control policy is the varieties brought from outside the country. Traditionally, such varieties have been 'illegally' introduced from neighbouring countries and are grown by farmers in border districts. However, some of those introductions became very popular and have found its way to other districts, thereby significantly contributing to national agricultural output, but never found its way onto the variety list, because it was not bred or tested in the country. The rice varieties Sharna, Minikit (from India), Pajam and Naizershail (from elsewhere) are such examples. The popularity of Indian jute varieties is a similar case. Every year, large amounts of seed are smuggled into the country and procured by local farmers, which seed frequently out-competes BADC's jute varieties, despite the fact that it bears no trademark or quality label.

In maize, by contrast, which was de-notified in 1993, the crop sector has seen more variety introductions lately than ever before, nearly all of them by the private sector. This has helped the national production to increase nearly tenfold in 6 years time to 60,000 ton in 1999.

A revealing perspective on the variety release policy is provided by a display of the economic crop sector relevance (see Table 3). By comparing the volume and value of the notified crops (excluding sugarcane) against the de-notified crops it shows that the de-notified crops represent only a small proportion of total agriculture in Bangladesh. In other words, variety release in more than 90% of the crop sector is still controlled. Variety registration procedures thus still pose a serious hindrance on the development of the seed industry.

SEED PRODUCTION AND SUPPLY

During the period 1974-1993 the amount of quality assured seed in the country significantly increased (see Figure 2). It may, however, not be correct to assign the improvement in the seed supply entirely to the effect of the regulatory framework. Rather, the rapid expansion of the country's seed production and processing infrastructure (Table 4), the appropriate use of facilities, and the strong internal quality control system of the seed-producing agency (BADC) created the impact.

It is obvious, however, that the regulatory framework helped in systematising and formalisation of the seed system in the country. In doing so, it helped farmers to obtain a greater awareness of quality seed and higher yielding varieties, which resulted in higher outputs.

The private sector was in 1990 confined nearly entirely to imported vegetable seeds. This changed drastically with the introduction of the Seed Policy. Private sector, with the help of numerous donor-related projects, started to locally produce and supply labelled seeds. Until recently, this supply concen-

TABLE 3. Volume and Value of the Seeds Used for Crop Production in Bangladesh

Crop Type	Name of the Crop	Quantity* of Seed Used (ton)	Value of the Seed (in million Taka)	Value in Mill US$
Notified	Rice	308730	4630.95	
	Wheat	72480	1087.20	
	Jute	4464	89.28	
	Potato	170500	2557.50	
Total		**556174**	**8364.85**	**US$ 154.91 mill**
Non-Notified	Oilseeds	8280	165.60	
	Pulses	25470	509.40	
	Vegetable	3000	150.00	
	Maize	500	25.01	
Total		**37250**	**850.00**	**US$ 15.74 mill**
Grand Total		**593424**	**9214.85**	**US$ 170.65 mill**
Percentage of Non-Notified Crops		**6.3%**	**9.2%**	

* Quantity is based on recommended seed rates, not seed use
(Adapted from: Report of the Task Force, 1995)

FIGURE 2. Public and Private Sector Seed Supply Development in Cereals and Potato

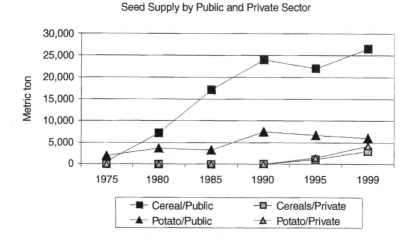

Seed Supply by Public and Private Sector

TABLE 4. The Development of the BADC Seed Processing Capacity

Year	1975	1980	1984	1990	2000
No. of Seed Processing Plants	1	6	14	16	21
Seed Processing Capacity (ton)	1,000	12,000	22,000	26,000	42,000

trated on more profitable crops, like vegetables, hybrid maize, and seed potato, but suppliers gradually made a move to the cereal seeds rice and wheat. Most of these so-called market-followers benefit from the weakness and shortage in the supply of BADC. Some NGOs involved in seed production thrive on secured markets among farmer member groups, while others enter markets that are not served by BADC.

Human Resource Development

Despite the Seed Policy only a dearth of qualified people in the field of seed technology is available in the country. Most skilled people are still employed in the public sector in the field of research, extension and seed production.

Facilities like seed stores, seed testing laboratories, production and processing machinery are mainly available with the public sector. These are hardly used for training private sector staff, and if so, mainly at the initiative of donor related projects.

Even though courses in seed technology at undergraduate and graduate level covering aspects of the seed industry are needed, few have been developed. Formal courses on seed technology, as foreseen in the Seed Policy, are yet to be established.

Private Sector Development

The private seed sector has made good progress since 1993. Assessment of the crop-wise impact, however, shows that most of the developments are found in a limited number of crops: rice hybrids, hybrid maize, seed-potato, and vegetables.

In the cereal sector, four companies have been licensed by the government to import and sell rice hybrid varieties. Other private sector organisations like the NGO's BRAC and GKF, and the company Novartis have started to supply cereal seeds of non-hybrid varieties nationwide, while smaller companies have established their own plants in cooperation with BADC, in anticipation that they can do a better job than BADC. In maize, a growing number of private suppliers have entered the market. Several companies have signed a license agreement with international seed companies to locally produce and market hybrid maize from imported parent lines. The production capacity is rapidly increasing. In seed potato eleven entrepreneurs are producing quality seed potato and market them. Domestic production of vegetable seeds has been introduced in the early 1990s and is a general practice now. Joint ventures with foreign seed companies have become a reality. Three such companies: East-West Seed Co., ACI, and McDonalds have been established to produce and market seeds in Bangladesh, some with own breeding capacity.

The Seed Policy emphasises a change in the role of public sector institutions towards active support of private sector development, which is evident in a number of cases. For example, private sector presently benefits from ready access to breeder's seed of publicly bred varieties and is able to obtain foundation seed from BADC at subsidised prices. BADC facilities are available to private sector against subsidised rates and utilisation is increasing. Such can be a significant service to start-up seed entrepreneurs. Under German assistance, a special private sector support unit has been established in the BADC to service small private seed companies, which program is gradually expanding. Services rendered directly and through technical assistance are in the field of seed technology, marketing, business planning, organisational development, information services and linkage to credit institutions.

However, implementation of the Seed Policy falls short in a number of areas. Entrepreneurs still find it difficult to obtain bank loans, either for industrial or working capital. While the government has declared seed production and marketing as an industry, no clear directives to seed industry matters have been provided to the financing institutions in the country. Custom duty on imported equipment for research, seed production, processing, storage and packaging materials are considered high which makes large scale production and marketing of seed expensive. Custom clearance procedures are still constraining imports. Value added tax is imposed on seeds if imported in bags, cans or in wrapped conditions, but is not levied if such seeds are imported in bulk. Such inconsistencies discourage import of quality seeds.

Monitoring and Coordination

At a time when the private sector initiatives takes momentum and are willing to invest in the national seed system, active support from the government is needed to guide the sector policy and regulations, make linkages with public sector organisations and remove obstacles. The Seed Wing of the MOA conducts this task in accordance with the outline envisaged in the Seed Policy (Table 5). Unfortunately, it appears that the Seed Wing is presently not well prepared to enact on this important task. After a number of staff changes, the section is limited in its resources and would need strengthening.

TABLE 5. Functions of the Seed Wing in the Ministry of Agriculture

Functions of Seed Wing (MoA) described in Seed Policy 1993
1 To help update policies and plan strategies for the development of the seed industry with special attention given to promoting private sector seed enterprises and to ensure implementation of such policies and strategies.
2 To monitor development and commercialisation of the seed sector.
3 To oversee and coordinate the production of breeder and foundation seed by public and private seed enterprises to meet farmers' demand.
4 To promote human resource development in the seed sector through training, seminars and workshops.
5 To develop a permanent cadre of trained and experienced seed technologists in public sector institutes to ensure sustained growth of the seed industry.
6 To plan and promote seed technology research in the NARS, BAU and the private sector.
7 To plan and implement a seed security system including maintenance of buffer stocks of seeds.

CONCLUSIONS

It cannot be denied that the new regulatory framework, outlined in the Seed Policy 1993, and implemented through the new Seed Act and Seed Rules has had a positive impact on the national seed system. The impact on the downstream seed supply is particularly evident, which was not feasible without making the seed certification voluntary. However, the upstream supply has not changed much and appears to be stagnant. Although fragile, the private sector, which was nearly absent a decade ago, is now present in virtually all crop seed sectors. Yet, when one compares the development with neighbouring countries like India and Thailand, the growth of the private sector is notably slow.

This evidently shows the following two facts.

First, the Seed Policy needs again to be reviewed and amended in those fields that ameliorate the upstream seed supply supporting the free flow of varieties. By retaining strict control on variety release of the major five crops the government has imposes controls on more than 90% of the seed business (see Table 3). This would not be so much of a problem if after implementing the Seed Policy significant additional funds were allocated to expand domestic breeding programs in order to compensate for the imposed restrictions on the free importation of varieties. While the ratio of research to agricultural gross value output in developed countries is about 5%, in many developing countries including Bangladesh it is less than 1%. In terms of investment the implications engaged with crop notification therefore have not or insufficiently been understood.

The authors are of the opinion that such high investment is neither feasible nor needed. There is a growing consensus worldwide that strict variety control is seriously constraining agriculture output and should, therefore, be removed (Gisselquist & Srivastava, 1997). Two solutions may be provided:

i.   The simplest solution might be to decide to step away from variety notification and allow free importation of varieties for all major and minor crops in the country. Such type of legislation is not uncommon and will be, for example, similar to the system implied in the USA.

ii. Alternatively, especially when such de-notification is meeting resistance, one could allow import of varieties from a number of countries, that have agro-ecological conditions similar to that in Bangladesh. This system of legislation has gained a wide acceptance, particularly in the countries of the European Union. Once a variety is released in one country, the other countries engaging in this scheme accept the variety on their variety lists without any further testing. Under Bangladesh conditions, this could imply that varieties released in West Bengal (India) may be allowed free import and trading in Bangladesh, so that farmers have ready access to seeds of these new varieties. A more fascinating example in this context may be suggested for potato, in which crop nearly all varieties originate from the Netherlands. One could justify that there is no harm into granting the free import to Bangladesh of any potato varieties recently released in that country provided proper quarantine procedures are followed.

Secondly, the implementation of the Seed Policy so far seems to have an unequal bias towards legal regulatory reform, but significantly lacks a sense of active promotion needed for seed industry development. As such, a lot of effort went into the development of the framework, such as Seed Act and Seed Rules, but the 'spirit' of the policy is not sufficiently carried forward. Lately, it appears that the controlling function has received more attention than the regulating functions as intended by the Seed Policy. Much of the early efforts to promote the Seed Policy were supported by the donor seed related projects, but with the number of projects drastically reduced, the Seed Policy intent seems to be gradually loosing intensity. Here lies a role for the coordinating authority to actively promote the ideas of the Seed Policy, link up with private sector and lobby with concerned decision-makers.

Policy decisions as mentioned above are urgently needed to improve the seed flow in the Bangladesh. Such is in the interest of the many small farmers who require strong innovations to improve cropping patterns and output.

REFERENCES

Bangladesh Government, 1993. *The Seed Policy of Bangladesh, Dhaka*. In: *The Bangladesh Gazette*, January 25, 1993 (Bengali version) and March 8, 1993 (English version).

Gisselquist D. & J. Srivastava, 1997. *Easing the Barriers to Movement of Plant Varieties for Agricultural Development*. Washington, DC, Worldbank Discussion Paper No. 367, p. 139.

Huda (1992). *Effect of Quality of Certified Seed and Farmers' Seed on the Productivity of Rice and Wheat* (unpublished PhD thesis). BAU, Mymensingh, Bangladesh.

Ministry of Agriculture, 1998. *The Seed Rules (1998)*. Notification by the Ministry of Agriculture, Government of the Peoples Republic of Bangladesh, Dhaka, 8 March, 1998.

Report of the Task Force (1995). *Requirements and Supply of Seed (Estimates for 1995-2000 and 2005)*. Ministry of Agriculture, Bangladesh Secretariat, June 1995.

Udin S.N., H. Smolders & N. Saleh (1995). Seed Quality of Secondary Food Crops in Indonesia. In: H. van Amstel, J.W.T. Bottema, M. Sidile & C. van Santen (Eds.) *Integrating Seed Systems for Annual Food Crops*. CGPRT No. 32. Proceedings of Workshop, Malang, East Java, Oct. 24-27, pp. 183-199.

World Bank (1973). *Appraisal Mission Report on Cereal Seed Project*, FAO/IBRD, Rome, pp. 1-23.

World Bank (1988). *A Study of the Role of the BADC in the Bangladesh Agriculture Development*, Refer seed operations in Vol. II. University of Illinois, Interpaks, Urbana, Illinois, USA.

World Bank & BCAS (1999). Bangladesh 2020; A Long-Run Perspective Study. In: *Bangladesh Development Series*, by World Bank & Bangladesh Centre for Advanced Studies. Dhaka, The University Press Limited, pp. 17-20.

Property Rights on Plant Varieties:
An Overview

Huib C. H. Ghijsen

SUMMARY. There is a general agreement that Intellectual Property Rights are important for promoting research and in particular practical plant breeding. Providing an effective IPR system by countries increases the access to foreign-bred varieties for their citizens. The design of such a national IPR system for plant varieties requires a careful balancing of exclusive rights and exemptions taking into account the current (informal) seed systems for the particular crops and their intended development.

The interest in Intellectual Property Rights for the protection of germplasm in the seed industry is growing with the increasing investment from the biotechnology as the value of the new traits will eventually be captured by outstanding varieties, which are bred by a combination of new technology and conventional breeding. The UPOV based Plant Variety Protection laws offer the most simple and adequate protection for breeders by combining a good scope of protection with a full freedom

Huib C. H. Ghijsen is affiliated with the Aventis Crop Science, Gent, Belgium.
Address correspondence to: Huib C. H. Ghijsen, Aventis Crop Science, Jozef Plateaustraat 22, B-9000 Gent, Belgium.

[Haworth co-indexing entry note]: "Property Rights on Plant Varieties: An Overview." Ghijsen, Huib C. H. Co-published simultaneously in *Journal of New Seeds* (Food Products Press, an imprint of The Haworth Press, Inc.) Vol. 4, No. 1/2, 2002, pp. 195-212; and: *Seed Policy, Legislation and Law: Widening a Narrow Focus* (ed: Niels P. Louwaars) Food Products Press, an imprint of The Haworth Press, Inc., 2002, pp. 195-212. Single or multiple copies of this article are available for a fee from The Haworth Document Delivery Service [1-800-HAWORTH 9:00 a.m. - 5:00 p.m. (EST). E-mail address: getinfo@haworthpressinc.com].

to operate. The Plant Patent in the USA, only accessible for asexually reproduced crops, except species with edible tubers, is of very little importance for the seed industry. The utility patent, also only in the USA available for conventional bred varieties, offers a very wide scope of protection at the choice of the applicant. This restricts the research freedom to the extent that allows only the study of the variety for making a phenotypic description. This very strong form of protection can be obtained without the "inventions" being tested against the prior art. As obtaining and defending utility patents is very expensive, it is only worthwhile for financial good yielding crops like corn and soybean. Furthermore, most of the patent claims have first to be tested in court before their validity will be known. Other forms of protection like trademarks, trade secrets and Material Transfer Agreements play a supplementary role, dependent of the crop and the circumstances. The agreement on Trade Related Aspects of Intellectual Property Rights (TRIPS) and the increasing use of IPRs have triggered some reluctance at the side of developing countries to implement western based IPR models like UPOV. Although UPOV is a well-balanced system for the protection of plant varieties, a solution must be found for the problem of the seed flow in the so-called informal seed systems. The Convention on Biological Diversity (CBD) and the International Treaty (IT) of the FAO, influence the development of IPRs on plant varieties, by requiring restrictions on IPRs and contributions to a fund for the maintenance of the agricultural genetic diversity. *[Article copies available for a fee from The Haworth Document Delivery Service: 1-800-HAWORTH. E-mail address: <getinfo@haworthpressinc. com> Website: <http://www.HaworthPress.com> © 2002 by The Haworth Press, Inc. All rights reserved.]*

KEYWORDS. Intellectual property, plant breeders' rights, patents, benefit sharing

INTRODUCTION

Since the beginning of the 20th century, increasing plant breeding activities by private companies called for protection of their investment in time and money. The first Plant Variety Protection (PVP) systems evolving in Europe were based on the developments in agriculture. Laws or regulations for seed certification, variety listing and eventually variety protection soon ruled plant breeding, seed production and seed trade. Germany and the Netherlands belonged to the first European countries introducing some form of PVP. The USA, however, was in 1930 the first country in the world to provide an intellectual property right for plant varieties, the Plant Patent Act. This system was and still is restricted to vegetatively reproduced crops, excluding plants producing edible tubers, like potato. The plant patent is similarly organised as the utility patent system, requiring a detailed description, photographs and drawings of the variety.

The European PVP systems use the central testing procedure for the technical requirements–Distinctness, Uniformity and Stability (DUS) as had been developed by the seed certification and variety listing authorities for the purpose of variety identification.

In 1961, after 4 years of negotiations, Belgium, Italy, Germany, the Netherlands and France signed the first Convention of the "Union de la Protection des Obtentions Végétales" (UPOV) in Paris. The countries, invited by the French government, considered at the beginning of their negotiations to adhere to the existing national utility patent laws based on the Paris treaty of 1883. For several reasons they decided eventually to develop a 'sui generis' system. The two most important of these reasons were the complexities of dependencies under utility patents–most varieties originate by crossing existing varieties–and the fact that the patent offices were not equipped to test the requirements for this kind of protectable subject matter.

The UPOV Convention has been changed in 1972, 1978 and 1991. The most important changes, with the view to strengthen the protection, have been made in the Convention of 1991. UPOV counts now (August 2001) 47 member states, of which 17 member states, including the USA, have ratified the UPOV 1991 Convention.

Utility patents have for a long time been considered unsuitable for protecting living matter. Even the application of patents in all agricultural activities was debatable. Most patent laws in the world nowadays forbid the patenting of plants, animals or varieties of plants and animals. The Diamond v. Chakrabarty case in the USA (see Box 1) meant an important milestone in the patenting of life debate. From then on, many utility patents have been granted in the field of biotechnology. The Hibberd case (see Box 1) extended the protective ability of the utility patent system to conventional bred plant varieties.

As a consequence, the utility patent became available in the USA as an interesting option for the protection of plant varieties. In this way, the USA offers uniquely three provisions to protect plant varieties: PVP for seed crops and edible tuber crops, Plant Patent for asexually reproduced crops, and the Utility Patent for all crops. Between the three systems there is neither any link nor any form of coherence or cooperation.

The UPOV Convention of 1978 bans double protection: varieties of a particular crop can be protected either by the UPOV system or by another system, not by UPOV and another system together. This ban has been lifted in the UPOV Convention of 1991 and consequently in the US-PVP act, modified in 1994 with the implementation of the UPOV 1991 Convention. Since this time, utility patents for plant varieties have been filed and granted in the USA on a

---

BOX 1

Diamond versus Chakrabarty, 206 USPQ 193 (Sup. Ct. 1980)

In this famous case the US Supreme Court, by five to four majority, ruled that a new strain of bacteria produced artificially (by bacterial recombination) was a patentable invention. (Until 1980 the USPTO refused claims to living systems as not being patentable subject matter.) The bacteria being able to disperse oil slicks were the product to be sold. It was therefore important to obtain a *per se* claim to the micro-organism. The Court interpreted the language of section 101 of the US patent Act which reads "whoever invents or discovers any new and useful process, machine, manufacture, or composition of matter, or any new and useful improvement thereof, may obtain a patent therefor," (..) to include "anything under the sun that is made by man." The relevant distinction was not between living and inanimate things, but between products of nature, whether living or not, and human made inventions. The Chakrabarty decision is being considered as *the* landmark in the patenting of living organisms, although the British patent for this invention had already been granted in 1976. Furthermore, the Paris Patent Treaty of 1883 provided since 1925 the possibility to patent inventions in the field of agriculture, including plants.

The Hibberd case (Board of patent appeals and interferences, Appeal No. 645-91, 1985) answered the question whether the PVP-Act and Plant Patent Act pre-empt the patenting of plants. Hibberd and cooperators had developed a tryptophane over-producing maize plant that offered improved nutritional value. The patent examiner's position had been, that the patentee's claims could not be granted because they were covered by the scope of the PVP-Act. Since the PVP-Act had been enacted later and is more specific than the Utility Patent Act in that it expressly applies to plants, the PVP-Act is the exclusive form of protection for plant life. The Board of appeals however, did not find any intent in the legislative history to restrict the scope of patentable subject matter and stated that a specific statute will not control a more general one in the absence of an irreconcilable conflict between them, nor would the overlap in coverage provide a basis for pre-emption. This decision apparently opened the gate for granting utility patents to plant varieties, although the Hibberd case concerned only plants with a specific trait and not varieties as such.

The Supreme Court of the United States accepted in February 2001 a case–J.E.M. Ag Supply *v.* Pioneer Hi-Bred International, in which a similar question concerning the pre-emption of the Utility Patent by the more specific PVP-Act is raised. This case may jeopardise the use of Utility Patents for the protection of conventional varieties.

---

regular basis. A number of these varieties have also been filed for the 'double' protection, both under the utility patent and the PVP Act.

In Europe the debate on the patentability of plants has been focussed for a number of years on the Novartis case. The final decision of the enlarged board of appeal of the European Patent Office in December 1999 ruled that varieties which contain a patented gene do not fall under the patent because of the prohibition of article 53 of the European Patent Convention (EPC) to patent plant varieties. This article was considered to avoid the accumulation of patents and

PVP. Plants, which contain the patented gene, however, will fall under the protection of the patent. This means in practice that the patent of a gene may include several varieties.

An important development in the last decade has been the coming into force of two international treaties. The agreement on Trade Related Aspects of Intellectual Property Rights (TRIPS-1994) of GATT obliges all WTO countries to provide for an effective Intellectual Property Rights (IPR) system for plant varieties before the year 2000 or 2005 (for least developed countries). This can be a system on its own (sui generis), such as UPOV, or patents or a combination thereof (TRIPS, article 27.3b). The Convention on Biological Diversity (CBD) of the 1993 Rio-Declaration recognises the sovereign rights of states over their biological resources. Access to biological resources can only occur with the prior and informed consent of states, and requires equitable sharing of revenues arising from the commercial use of these biological resources. The CBD stipulates that IPRs must not conflict with the conservation and sustainable use of biodiversity. As most WTO countries (but not the USA) are signatories of the CBD, these two treaties embody the conflict between public and private interests with regard to biological resources. In this chapter some attention will be paid to the practical consequences of these treaties.

As the most commonly used (sui generis) systems are based on the 1978 and 1991 UPOV Conventions, an important part of this chapter will be dedicated to UPOV. Secondly, attention will be paid to current use of the utility patents in the USA, as this development has some contradictory elements. Finally, some words will be spent on the supplementary role of trademarks in the protection varieties in some species and the use of Trade Secrets and the Material Transfer Agreement.

REQUIREMENTS, EXAMINATION AND GRANTING OF PROTECTION

Plant Variety Protection

Plant Variety Protection also known as Plant Breeders' Rights is granted to a plant variety if the variety is New and Distinct, Uniform and Stable (DUS).

A disclosure of the variety, as required by the utility patent is not foreseen in the PVP system. Most varieties, however, do appear on the market, except the parent lines of hybrid varieties. Parent line varieties can even remain secret while they are protected.

Novelty

UPOV 1991 requires that material of the variety must not, before the date of application, have been traded more than one year within the country of application or four to six years outside this country. Under UPOV 1978 the one year 'grace' period is optional, but all UPOV 1978 based national legislation require the PVP application to be filed before trading or even before offering the variety for sale.

Distinctness

This is mainly based on a clear visible phenotypic difference between the new variety and the most similar existing varieties that are commonly known in trade or still under test. Distinctness is a relative issue: it is valid for the particular location, the period tested and the reference varieties with which it is compared. The interactions often observed between varieties, locations, observers and years are typical for living organisms, which make variety testing complicated but also challenging and interesting. It is fundamentally different from non-living matter like cars, computers and chemicals: these items can be measured by anyone at any time at any location, always yielding the same results. The use of molecular fingerprinting in DUS testing was expected to resolve the problem of variety/environment interaction. These techniques are however not yet sufficiently robust and cannot totally replace the phenotypic traits. Furthermore, these techniques are, due to the high costs, only interesting in a few big crops and finally raise the important problem of a reduced variety distance, resulting in a decreased scope of protection. The molecular fingerprinting techniques may play an important role in the management of large reference collections by identifying the varieties most similar to the new variety and growing them side by side in field trials. The new 1991 UPOV Convention requires only a clear difference in the expression of at least one characteristic, while UPOV 1978 offers a somewhat wider scope by requiring a difference of one or more important characteristics.

Uniformity

Uniformity is always assessed in relation to the way of propagation. Vegetatively propagated and selfed varieties have the highest uniformity requirement with a tolerance of 1% off-types on average. Seed certification requirements are often more strict with a tolerance of only 0.1%.

The UPOV uniformity requirement has yet met much criticism from various sides, mainly developing countries, who fear that the broad adaptation of existing varieties (landraces) might be endangered by the development of strict uniform varieties with a narrow genetic basis. According to its variety definition however, UPOV 1991 acknowledges the existence of broad varieties. Article 1 (vi) says: "variety" means a plant grouping (...), irrespective of whether the conditions for the grant of a breeder's right are fully met (...)." An existing variety not sufficiently uniform for protection is, therefore, regarded as a public variety of common knowledge, which has to be taken into account by establishing the distinctness of new varieties. Furthermore, protecting broad, non-uniform varieties, would greatly enlarge the scope of protection of such a variety and would complicate the task of its description, to maintain its stability and to defend its protection. Finally, if particular characteristics, like disease resistance, need to be heterogeneous in order to be effective under field circumstances, such characteristics may simply be left out of the DUS testing.

Stability

Stability is inherent to the system: if a variety changes significantly in the expression of one or more of its relevant characteristics with respect to its original description and its seed or plant material sample, it has become a different variety, which is not protected. This stability requirement may become important if a PVP must be defended in an infringement case. Systematic testing of stability for PVP, as required in some national and supranational PVP laws, is not necessary and a waste of resources.

For a different and public purpose, in order to provide the consumer with seed of the true variety, seed lots are systematically checked whether they belong to the claimed variety.

In the European UPOV-member countries all new varieties are tested in official grow-out tests to verify the DUS of seed crops for listing, admittance to seed trade and granting of PVP. This has resulted in large reference collections of existing varieties, which are compared in field trials each year. Although this is a rather costly affair, it has shown advantages like a high efficacy, a strong accumulation of technical and variety knowledge and an effective weapon against infringement, especially in seed crops. On the basis of the accumulated technical knowledge, UPOV has developed a set of technical guidelines for the DUS testing of crops, with the aim of harmonising and supporting the technical testing. These guidelines are being designed and discussed in the yearly meetings of crop specific technical working parties, which are open to the professional organisations.

In the USA, Latin America, Australia and Canada the breeders need to provide a detailed variety description to the PVP-office to perform the DUS-testing. These descriptions are tested for distinctness in a computer database. In Australia, the breeder's trials are supervised by a 'qualified person' selected by the PVP-office. The uniformity and stability are declared by the breeder to be satisfactory. If this is not correct or not true, the sanction is, that the PVP title can be annulled by anybody during the whole lifetime of the PVP. Nullification has the effect that the PVP has never existed. Nullification is possible for varieties tested in any system, if the variety was not new or distinct at the moment of granting the protection.

After a crop depending testing period of 1-3 years a title of protection is granted for a minimum period of 15-18 years (UPOV 1978) or 20-25 years (UPOV 1991) *from the date of granting:* 18-25 years for vines and trees, 15-20 years for all other species. Some national laws provide a longer protection of 25-30 years.

Plant Patent

An asexually reproduced plant, except edible tuber plants, may obtain a plant patent in the USA if the plant was:

- Invented or discovered,
- New,
- Distinguishable,
- Non-obvious.

Invented or Discovered

A discovery, the 'finding' of a plant must take place in a cultivated area (e.g., a mutant or crossing) and not in the wild nature.

New

The plant variety must not have been sold, released or publicised in the USA more than one year prior to the date of application.

Distinguishable

The plant must have shown to differ from known, related plants by at least one distinguishing characteristic.

Non-Obvious

The invention must be not obvious to a person skilled in the art. The invention is obvious if it can be shown that it could be reasonably expected by the references relied on by the examiner. Examples of breeding methods resulting in obvious varieties are the use of mutagens in doses known to work, tissue culture of diverse chimera to separate the genotypes and the application of colchicine to double the chromosome number.

With the application, a description of relevant prior art and a full and detailed botanical description of the invention are required, accompanied by photographs drawings and a claim to the plant as a whole. The plant patent is limited to one claim–different from the utility patent where numerous claims are allowed–which may make reference to specific characteristics of the plant, but not to parts or products. The description of the claimed plant will be compared with the closest available prior art. The applicant may provide supplementary data to distinguish his plant invention from close prior art.

The plant patent is granted for the period of 20 years *from the date of filing*.

Utility Patent

Utility Patents are granted in respect of products or processes, which are:

- New,
- Inventive or non-obvious,
- Useful or capable of industrial application.

New

An invention must be new in respect to the 'state of the art,' which is everything, world-wide, made available to the public by description or use or any other way. Inventions are usually compared with existing publications in international accessible databases. In the USA such a database for plant varieties with full descriptions is available at the US-PVP office in Beltsville but this is not being used for testing the newness of utility patent applications of plant varieties.

Useful

Having a practical purpose in any kind of industry, including agriculture. Methods of medical treatment for instance are defined as being incapable of industrial application. This requirement is marginally tested: usefulness is assumed present unless the opposite is clear.

Inventive

Inventive or not obvious to a person skilled in the art with respect to the state of the art. In utility patents, this is a difficult and vague criterion, predominantly subject to the judgement of the patent examiner, which results in office to office differences in the granting of patents. The utility patent examiner is considering plant varieties as the 'unexpected' and therefore non-obvious result of a systematic breeding process. This statement is true for most plant varieties, as no breeder is able to predict precisely the result of his crossing and selection work. In this way, the inventiveness requirement has become a dead letter.

The patent application procedure requires disclosure of the invention by a full description. This description is published 18 months after application (the

USA follows European practice in this respect since late 2000). This description must be in such a way that it enables a person skilled in the art to reproduce the invention. Soon after patents for living organisms were being granted it was recognised that it is virtually impossible to reproduce a bacterial strain or plant variety from a written description. To overcome this problem a living sample is deposited in a recognised and public accessible culture collection or seed bank. The majority of the developed countries have adopted the Budapest treaty of 1977, which rules the formalities of the deposit and the maintenance of the samples.

As has been mentioned before: this does not count for the PVP system. Seed of varieties protected by PVP is not systematically being stored in a public accessible seed bank.

The Claims

An important part of the patent application is the formulation of the claims, which define in fact the scope of protection. It is this part which make utility patents fundamentally different from PVP. This topic will further be discussed in the next paragraph dealing with the scope of protection of PVP and patents.

The European patent system knows a procedure for any interested party to file an opposition within 9 months after publication. The advantage is the reduction of granting 'bad patents,' which can only be revoked by a very expensive juridical procedure. The disadvantage of the European opposition procedure for the patentee is the stretching of the period before the patent will be granted, which can take several years.

Utility patents are in force from the day of granting until 20 years after the date of filing, except for pharmaceuticals, for which the period can be extended to 25 years due to long testing time for marketing approval.

SCOPE OF PROTECTION

The Scope of Plant Variety Protection

The scope of protection for PVP is very well defined and rather limited, although there is a great difference between UPOV 1978 and 1991. The original UPOV scope of protection concerned only the traded reproductive material. Farmers may under UPOV 1978 freely use their harvested grain to seed the next crop. This is known as the farmer's privilege. Even the exchange of seeds between farmers was allowed under the former UPOV 1978 based US-PVP act by the provision of a crop exemption. The 1991 UPOV Convention has extended the scope of protection by including the following items in the exclusive rights of the breeder:

- all handling with all material produced of the protected variety.
- the harvested product of the protected variety.
- all handling with all material produced and the harvested product of varieties essentially derived from the protected variety and varieties that need the repeated use of the protected variety (parent lines).

The extension of the protection to all material produced, irrespective whether it is for trade or own use, means that the general farmer's privilege has been abolished. It has been replaced by a restricted farmer's privilege, which will be discussed in the next paragraph.

The harvested product provision is of special interest, e.g., the cut flower industry to capture the import of flowers produced in countries where no PVP exists. The notion of essential derivation originates from the mutant problems in the field of ornamentals and the concern of breeding companies in the 1980s, that the new technique of genetic engineering would easily circumvent the narrow UPOV protection. The definition of essential derivation, however, is rather vague and complicated and susceptible to different interpretations. Summarised it says that a variety is essentially derived from an initial variety if it is predominantly derived from that initial variety, is distinct and remains unchanged in the expression of the essential characteristics, except for the differences caused by the act of derivation. Examples of essential derivation are the selection of a mutant, a (somaclonal) variant, genetic engineering and back crossing. The discussion within the crop sections of the world wide breeder's organisation ASSINSEL focuses on the genetic distance between varieties, measured by DNA fingerprinting techniques. The aim is to agree on thresholds per crop beyond which a variety is to be considered essentially derived. The discussion among the maize breeders is most advanced with threshold proposals of 80-85% genetic conformity, which means that the owner of an initial variety can claim a large area of the genetic variation. This development means a break with the UPOV tradition of accurately defining the scope of protection and a step back to the complicated jumble of dependencies, which it avoided wisely in 1961. Leaving the setting of these thresholds to the seed industry means that the seed industry itself determines now an important part of the scope of protection. The decision whether a variety is essentially derived from another initial variety has great consequences: the protection of the initial variety stretches over the derived variety. The possibility for the owner of an essentially derived variety to obtain a compulsory license from the breeder of the initial variety is virtually impossible in the PVP system. In contrast to the initial variety, the protection of the derived variety confines only to the variety itself. This means that it is also important to determine whether a claimed initial variety is a true initial variety. In this respect the threshold as established for essentially derived varieties influences also the position of the initial varieties:

the lower the threshold, the wider the scope of protection and the greater the chance that a variety, claimed to be an initial variety, may eventually appear to be essentially derived from another, older and possibly non-protected variety.

Theoretically the thresholds can be challenged in a civil court case, but due to the complicated nature of the matter and the high litigation costs involved, this is not likely to happen, except possibly by big companies who can afford these costs. Litigation concerning PVP issues has been rare between companies, partly due to the work of the UPOV bodies–Council, Legal and Administrative Committee and Technical Committee–in which many technical and legal matters are regularly discussed with a considerable contribution from the breeder's organisations.

The Scope of Protection of Plant Patents

The plant patent grants the right to exclude others from asexually reproducing the plant and from using, offering for sale, or selling the plant so reproduced or any of its parts. This formulation by paragraph 163 of the US Patent Act restricts the scope of protection to the plant variety itself and leaves room for closely resembling varieties (mutants) to be protected independently from an initial variety, bred by crossing and selection. This means that the breeders of ornamental and fruit varieties for which PVP in the USA is not provided, do not enjoy the PVP extension to essentially derived varieties. This provision of essential derivation is especially important for ornamental and fruit breeders due to the easy discovery of mutants by the growers, who observe thousands of plants in their production fields.

The Scope of Protection of Utility Patents

The scope of the utility patent is determined by the formulation of the claims. The claims may, as in the case of the plant patent, be restricted to the plant and its parts. Usually, however, the patentee tries to formulate his claims as broad as possible (see Box 2).

It is the task of the patent office to examine these claims and revoke the ones that cover the prior art or are, to the discretion of the patent examiner, far beyond the scope of the invention. The applicant may appeal to decisions of the examiner. In Europe other parties may file an opposition to patents and its claims, which is not possible in the USA. With the absence of any opposition and plant breeding expertise, the US Patent and Trademark Office (USPTO) grants apparently any claim filed by the applicant for a utility patent on a conventional plant variety.

The European Patent Office on the other hand does not grant patents for plant varieties as this is prohibited by article 53(b) of the European Patent Convention (1973). This position has been clearly reconfirmed in the Novartis de-

---

BOX 2

Example of broad patent claim. Claim 23 of US patent 6,096,953 granted 01-08-2000

"A maize plant, or parts thereof, wherein at least one ancestor of said maize plant is the maize plant of claim 2, said maize plant expressing a combination of at least two traits of this plant selected from the group consisting of: a relative maturity of approximately 108 to 115 based on the Comparative Relative Maturity Rating System for harvest moisture of grain, high yield, above average cold test results, above average plant height, above average ear height, above average early stand count, fewer than average barren plants, above average root lodging resistance, and adapted to the Central Corn Belt and Southeast regions of the United States."

This means that all plants resulting from a cross with the patented plant, irrespective the number of generations of further crossing and selection that show a combination of two very general, agricultural characteristics, do fall under this patent.

---

cision of the enlarged Board of Appeal of the European Patent Office (Case G0001/98-decision of 20-12-1999): "the exception to patentability in article 53(b), 1st half sentence, EPC applies to plant varieties irrespective of the way in which they were produced; a claim (of a product or process, for instance the insertion of a gene-HG) wherein specific plant varieties are not individually claimed is not excluded from patentability under article 53(b) EPC, even though it might embrace plant varieties." The motivation to maintain the prohibition of patenting plant varieties is based on the availability of the specific UPOV based PVP acts at the national and supranational level in the EPC member states. Although this may be regarded as a formal and not a substantial reason it has the advantage that the protection of plant varieties is dedicated to a legislation of its own kind (sui generis) that is specifically tailored for this purpose.

The wide scope of protection of plant varieties in the US Utility Patent system might be challenged in court, but this is a procedure too costly for the average plant breeder, who would rather circumvent these patents than risking an infringement procedure.

EXEMPTIONS

Exemptions Under Plant Variety Protection

The Breeder's Exemption

The UPOV Convention of 1961 introduced a very important exemption to the exclusive rights of the breeder, the so called 'breeder's exemption.' This implies that breeders may unrestrictedly make use of protected varieties for breeding new, commercial varieties. The backgrounds of this provision was the existing habit of breeders to build upon the best available varieties present, the notion that most varieties were as a result interrelated and that claiming further breeding as an exclusive right would lead to a not-to-unravel jumble of relations between varieties. From a society point of view it bore the advantage of competition and the avoiding of monopolisation of specific breeding goals. For successful companies it has sometimes the disadvantage of closely being followed by 'me too' varieties, although the marketing position of the first product is mostly much better than the remakes. For the problem of the very close, almost identical varieties, the notion of essential derivation has been introduced as discussed in the former paragraph.

The Farmer's Privilege

A much discussed exception is the farmer's privilege: notwithstanding the exclusive rights of the breeder the national legislator may, for certain crops (in practice agricultural food crops), allow the farmer to use for his own farm the harvest from his own farm for the seeding of the next crop, without the consent of the breeder, but taking into account the legitimate interest of the breeder. The last part has resulted in different implementations in the PVP laws: from paying a substantial royalty for the 'farm saved seed'–in the EU–to paying no royalty at all–in the USA and for small farmers in the EU. The royalties on the farm saved seed in the EU are to be collected on the level of the member state by private organisations. Until now this is only quite successfully implemented in the UK, Sweden, the Netherlands, and Germany. Farmer's resistance is quite strong in some, most southern European countries, where the seed saving in cereals can be as much as 90% of the total seed market.

The habit of farmers to save, exchange and produce seed for their families and neighbours is a millennia old habit. It is fully understandable that breeders need some return on their investments, but it creates quite some resistance if the enforcing of intellectual property rights thwarts these old habits. UPOV did not sufficiently acknowledge the political sensitivity of this issue while drafting the 1991 Convention. Especially not in the view of the developments following the TRIPS agreement, which obliges all WTO member states, under which the majority is formed by developing countries, to implement an intellectual property rights system for plant varieties. UPOV 1978 gives more freedom to the national legislator to provide solutions for political sensitive issues than UPOV 1991. It is regrettable that developing countries cannot become a member of UPOV under the 1978 Convention anymore, but are obliged to follow the 1991 Convention, which is more adapted to the developed countries. This is in contrast with the fact that the present UPOV member states still adhering to the 1978 Convention can maintain their 1978 based membership without restrictions. It would have been better, as is mostly used in international treaties, to add a more stringent protocol to the existing, more lenient 1978 Convention. This would have made it easier for important developing countries like India, Thailand and the Philippines to join the internationally harmonised UPOV system. Presently these countries have PVP bills which allow seed exchange between farmers and acknowledge Plant Breeder's Rights of farmers and communities on old landraces, provisions which are not acceptable under UPOV 1991. (India may still accede under the UPOV 1978 Convention.)

The use of plant material for private, non-commercial purposes is free. Farmers may for instance always re-use seed of a protected variety to produce grain for their own feed. This freedom is important for small, subsistence farmers in developing countries.

Exemptions Under Plant Patent

According to the scope of protection as formulated in paragraph 163 of the Plant Patent Act, commercial breeding is allowed but production of material by growers for their own use is prohibited.

Exemptions Under Utility Patent

In the USA the mere use of a patented product usually constitutes infringement under Section 271(a) of the Patent Act: "except as is otherwise provided in this title, whoever without authority makes, uses offers to sell, or sells any patented invention within the United States or imports into the United States any patented invention during the term of the patent therefor, infringes the patent." However, there is a line of authority suggesting that "use of a patented product for non-commercial, experimental purposes is not an act of infringement." This experimental use doctrine is especially germane to the pharmaceutical industry. The Drug Price Competition and Patent Term Restoration Act expressly states that gathering data in preparation for the registration of drugs does not constitute patent infringement. It is assumed that in the case of plant breeding, the testing and describing of the patented variety is allowed although DNA fingerprinting may be claimed by the patentee as an exclusive right. Crossing with the patented variety is hard to control by the patentee, but the commercialisation of such a variety is an infringement of the patent. With the aid of DNA fingerprinting infringements can nowadays be detected more easily and most maize companies for instance analyse all new varieties that enter the market.

The national European patent laws formulate the research exemption explicitly, but do not allow, like in the USA, the pre-testing of generic drugs by competitors of the patentee before the patent has expired. In the case of varieties falling under the patent of a gene, it is assumed that experimenting (crossing) with the plants containing the gene is permitted, but not the unauthorised commercialisation of varieties in which the gene is present. It is questionable whether testing for listing and PVP is permissible without the consent of the patentee.

OTHER RELEVANT (INTELLECTUAL) PROPERTY RIGHTS SYSTEMS

Trade Secrets

In the USA some seed companies use State laws concerning trade secrets to protect the inbred lines of hybrid varieties. The reasoning is that parent lines, as long as they are not being released in public, are to be considered as the companies' trade secrets. Labels on hybrid seed bags warn the user that the seed may not be used for 'parent line fishing.' As litigation about this issue still continues, it is not clear whether this form of protection will become important. The situation concerning the position of parent lines is generally unclear, also within the UPOV system. It comes down to the question whether the hybrid seed is to be regarded as (harvested) material of the female line. If so, the commercialisation of the hybrid seed damages the novelty of the female line and constitutes it as part of the public knowledge or prior art.

Trademarks

Trademarks are mainly used in the cut flower industry although some seed companies may use them. In the case of import of unauthorised production, which is mostly marketed under a well-known trademark, the owner of the variety can easily confiscate the flowers. Although the variety name itself must under UPOV laws be a generic name, it is allowed to add a trademark. Generally growers and consumer better know trademarks than the official variety name. The protection of a trademark can last forever as long as the fees are being paid. This means that the protection of a trade marked variety can be extended after the expiration of the PVP or Patent. In theory, the no longer protected variety may freely be traded under a different trademark or its generic name, but these are unknown to the public. PVP remains important in combination with trademarks. Infringers could otherwise easily from the start produce and strongly market the non-protected variety under a different trademark and possibly also under a different generic name.

Material Transfer Agreement (MTA)

This kind of contract is increasingly used in obtaining and exchanging germplasm. The MTA may contain conditions, which restrict the receiver of the material to use it for research purposes only and explicitly not for further breeding. The legal validity of such an 'agreement' is questionable if a utility patent does not protect the variety. Research includes crossing and the commercialisation of resulting products is not forbidden by PVP. On the other hand companies, apart from some exceptions, do not like litigation which burns too much of their resources. In this way MTAs do serve their purpose by putting up some kind of barrier. This is also true for a kind of one sided, 'sui generis' MTA, put on the seed bags or labels, which prohibits the user of the seed to re-use the harvest or use it for plant breeding purposes.

The FAO International Undertaking on Genetic Resources for Agriculture and the Convention on Biological Diversity

This politically important and complicated issue is discussed here as it is directly influencing the development of intellectual property rights for plant varieties in the world.

The FAO International Undertaking has formulated the notion of farmer's rights (not to be confused with the 'farmer's privilege') in 1989 as the rights of farmers, especially in developing countries, to benefit from their contributions in conserving the agrobiodiversity in the past, the present and the future. "Farmers' Rights" will be implemented through international funding on plant genetic resources, which will support plant genetic conservation and utilisation programmes. The discussion is furthermore connected with the national sovereignty of genetic resources as formulated by the CBD and embodied in a Prior Informed Consent for companies to collect germ plasm in-mostly-developing countries. Because obtaining such a consent through bilateral negotiations is far more expensive in terms of manpower for each individual organisation than the fee to be paid and countries of genetic origin are sometimes difficult to identify, a more general solution has been sought. This has resulted in negotiations in a contact group of the FAO Commission on Genetic Resources for Food and Agriculture about a multilateral agreement in which the species that are freely accessible for plant breeders will be listed. This free access is in exchange for the establishment of a genetic resources fund. Proposals include the provision of a voluntary donation to this fund in the case varieties derived from genetic resources are protected by a form of IPR that does not restrict further plant breeding (e.g., PVP) and an obligatory donation if the IPR restricts further breeding (e.g., utility patent). Another important topic is the protection by IPRs of genetic resources as such, "in the form received." The discussion focuses on the point whether genes isolated from a seed sample, obtained from a genetic resource, is patentable.

The negotiations were concluded in November 2001 with the adoption of the International Treaty on Plant Genetic Resources for Food and Agriculture (www.fao.org/ag/cgrfa).

The Treaty contains a list of 65 quera and species that fall under the 'Multi-lateral Systems' for access and benefit sharing. For this list, containing almost all major food crops, access to genetic resources is significantly facilitated. For all other crops, bilateral contracts have to be concluded stating the conditions for access (with special emphasis on 'Prior Informed Conset') and arrangements for the sharing of benefits arising from the use of the resources among the contracting parties.

Regulating Genetically-Modified Seeds in Emerging Economies

Patricia L. Traynor
John Komen

SUMMARY. This article focuses on the implications of biosafety regulation on national seed policy. Biosafety regulation–the policies and procedures adopted to ensure the environmentally safe application of modern biotechnology, in particular, the release of genetically-modified organisms–has been extensively discussed at various national and international fora. Much of the discussion has focused on developing guidelines, appropriate legal frameworks, and, at the international level, on developing a legally binding International Biosafety Protocol. A major challenge in emerging economies will be to build the necessary infrastructure and human capacities to implement international and national biosafety guidelines or laws. Biosafety regulation adds a new dimension to national seed policies, which now have to consider their role in the process leading to the commercial release of genetically-modified varieties, including environmental assessments. In addition to biosafety review and approval, GMO crops intended for commercial use also are subject to existing regulations for seed registration, and may involve sanitary and phytosanitary regulations governing import, as well. There is potential for the respective authorities to overlap, which could lead to discord between the agencies and confusion on the part of the proponent. Capacity building to deal with the development of, and trade in genetically-modi-

Patricia L. Traynor is affiliated with Virginia Polytechnic and State University, USA.

John Komen is affiliated with International Service for National Agricultural Research, ISNAR 93375, 2509AJ The Hague, The Netherlands.

Address correspondence to: Patricia L. Traynor, Virginia Polytechnic Institute and State University, Blacksburg, VA 240614 USA.

[Haworth co-indexing entry note]: "Regulating Genetically-Modified Seeds in Emerging Economies." Traynor, Patricia L., and John Komen. Co-published simultaneously in *Journal of New Seeds* (Food Products Press, an imprint of The Haworth Press, Inc.) Vol. 4, No. 1/2, 2002, pp. 213-229; and: *Seed Policy, Legislation and Law: Widening a Narrow Focus* (ed: Niels P. Louwaars) Food Products Press, an imprint of The Haworth Press, Inc., 2002, pp. 213-229. Single or multiple copies of this article are available for a fee from The Haworth Document Delivery Service [1-800-HAWORTH 9:00 a.m. - 5:00 p.m. (EST). E-mail address: getinfo@haworthpressinc.com].

fied seeds presents a challenge to "emerging economies" where regulatory agencies are often fragile and vulnerable to political change. *[Article copies available for a fee from The Haworth Document Delivery Service: 1-800-HAWORTH. E-mail address: <getinfo@haworthpressinc.com> Website: <http://www.HaworthPress.com> © 2002 by The Haworth Press, Inc. All rights reserved.]*

KEYWORDS. Biotechnology, genetic modification, biosafety, regulation, risk assessment, capacity building, Egypt, Argentina

INTRODUCTION

Considerable public debate has emerged over the perceived benefits and risks of genetically modified organisms (GMOs), which in many countries is leading to increased government regulation of R&D and trade in GMOs. This has added a new dimension to national seed policies, which now have to consider their role in the process leading to the commercial release of genetically modified varieties, including environmental assessments. Capacity building to deal with the development of, and trade in genetically modified seeds presents a challenge to "emerging economies" where regulatory agencies are often fragile and vulnerable to political change.

"Biosafety" is associated with the use of genetically-modified organisms (GMOs) and, more generally, with the introduction of non-indigenous species into natural or managed ecosystems. Biosafety regulation–the policies and procedures adopted to ensure the environmentally safe application of modern biotechnology–has been extensively discussed at various national and international fora. Much of the discussion has focused on developing guidelines, appropriate legal frameworks, and, at the international level, on developing a legally binding International Biosafety Protocol.

At the national level, a major challenge in emerging economies will be to build the necessary infrastructure and human capacities to implement international and national biosafety guidelines or laws. Establishing and maintaining a functional, effective biosafety system presents various challenges. It requires adequate and dependable funding. It entails education and coordination across government ministries, universities and research institutes, private sector interests, individual scientists, and the public. Significant investments may be needed in training and human resource development, information and communications systems, and laboratory and greenhouse facilities.

Recent research initiated by the International Service for National Agricultural Research (ISNAR) analyzes the functioning, and possible enhancement, of national biosafety systems. Case-study research has been completed in

Egypt and Argentina. This article builds on the research results to date, focusing on the implications of biosafety regulation on national seed regulation.

GMOs–What Are the Risks?

Not surprisingly, there are ongoing discussions over what constitutes an environmental risk and what is the limit for acceptable risk. Currently, biosafety reviews generally focus on a limited number of environmental issues associated with the release of transgenic crops. Two of these concern the possibility that crops or their relatives may invade new territory, displace existing plant communities, or reduce species biodiversity. They may have added importance in regions that are centers of origin or diversity for the crop.

Weediness–the potential for a crop to become established and to persist and spread into new habitats as a result of newly introduced genes–is an issue when there is scientific evidence that acquisition of the new genes is sufficient to convert a domesticated species into a successful weed. *Gene flow*–in which new genes are spread by normal outcrossing to wild or weedy relatives of the engineered crop–becomes an issue if the new trait(s) confers a fitness advantage and becomes stably introgressed into the recipient genome. *Toxicity* is an issue associated with human health concerns over allergenicity and the safety of biotechnology foods and potential negative effects on nontarget organisms, especially beneficial species. *Pest and pathogen effects* include a range of possible consequences such as the generation of novel viruses by molecular exchange within a transgenic plant, or emergence of target pest populations resistant to an engineered control mechanism, such as the expression of Bt toxins.

An International Biosafety Protocol: Implications for Seed Trade

In January 2000, over 130 governments reached agreement on the International Biosafety Protocol, which will regulate the safe transfer, handling and use of *living modified organisms* (LMOs)[1] resulting from modern biotechnology. The ultimate goal of the agreement is to ensure an adequate level of protection to potential adverse effects from LMOs on the conservation and sustainable use of biological diversity, taking also into account their possible risks to human health. To date, around 80 countries have signed the protocol, while two countries have ratified. The protocol will enter into force once 50 countries have ratified it.

The two cornerstones of the Protocol are the concepts of *Advance Informed Agreement* (AIA) and the *Precautionary Principle*. Through the Protocol, AIA enables an importing country to subject all first imports of LMOs to risk assessment before taking a final decision on import. The Protocol provides de-

---

1. Living modified organisms (LMOs) are organisms whose genetic material has been altered through modern biotechnology and which are capable of propagation.

tails on the whole process of notification, acknowledgment and decision, which is supposed to be completed in 270 days. Detailed information will have to be provided by the importer on notification and LMOs should be clearly identified by accompanying documentation. In addition, the precautionary principle, as applied in article 11 of the Protocol, asserts that

> Lack of scientific certainty due to insufficient relevant scientific information and knowledge regarding the extent of the potential adverse effects of a living modified organism on the conservation and sustainable use of biological diversity in the Party of import, taking also into account risks to human health, shall not prevent that Party from taking a decision, as appropriate, with regard to the import of that living modified organism intended for direct use as food or feed, or for processing, in order to avoid or minimize such potential adverse effects.

In plain terms, including this principle allows countries to block imports of seeds of genetically modified plant varieties on a precautionary basis even in the absence of sufficient scientific evidence of their harmfulness. The protocol does not apply to processed foods derived from LMOs. For bulk commodities containing an LMO component, documentation will have to state that the shipment "may contain" LMOs and that the contents of the shipment are not intended for planting.

Consequently, the most immediate impact of the protocol on agriculture will be on import and export of seeds intended for planting. Before a GM seed can be shipped for the first time, the importing country must decide whether or not to approve it. If the seeds are approved for import, they will need documentation provided by the exporter specifying their identity and traits. Under the protocol, a 2-year process was established through which further documentation requirements will be considered.

BUILDING NATIONAL BIOSAFETY SYSTEMS

Biosafety developments at the national level are not only spurred by discussions in international fora. In addition, there are a number of "push" factors for governments to have a functioning biosafety system in place, such as:

- The number of genetically-modified products approved for commercial release is steadily increasing in industrialized countries, and these are spreading to developing countries and countries with economies in transition;
- The number of institutions that have the capacity to apply modern biotechnology is expanding, particularly in emerging economies;

• A range of international organizations and research institutes are developing or transferring to developing countries agricultural biotechnology products and tools which need biosafety review.

To date, however, only a few developing countries have established functional biosafety review systems. The preamble to the Biosafety Protocol recognizes the "limited capabilities of many countries, particularly developing countries, to cope with the nature and scale of known and potential risks associated with living modified organisms."

Although the Protocol provides rudimentary guidance for biosafety decision-making, risk assessment and risk management, there is a tremendous challenge for most signatory countries to implement the agreement nationally.

Appropriate oversight of LMO releases in the environment is commonly achieved through the establishment of a national biosafety system. Four elements need to be in place–the policy, the people, the process and the public ("the four P's")–that together help to generate environmentally responsible decisions. The four P's are described below.

The Guiding Policy

Policies for biosafety oversight define the objectives for regulating GMOs and the scope of regulation. Guidelines for implementation should clearly articulate what is subject to biosafety review and what is not. Do they apply to products of all modern biotechnologies or only to rDNA research? Are laboratory and greenhouse experiments included, or only releases into the environment? Do they apply to recombinant vaccines and pharmaceuticals or non-agricultural applications such as industrial fermentation technologies? Are there provisions for commercializing products? How will food safety be evaluated? Which Ministries are to be involved?

Biosafety regulations may be in the form of new legislation, adaptation of existing regulations, or non-legislative guidelines issued, for example, by Ministerial Decree. Such documents typically authorize the formation of national and institutional Biosafety Review Committees, specify their respective duties and membership, and describe application and review procedures for environmental releases of GMOs. In deciding which form to use, the merits and drawbacks of each option must be weighed. For example, while a new law carries the weight and enforcement power of government regulatory authorities, it is difficult to amend, and a lengthy effort is needed to enact and implement it. A ministerial decree, in contrast, is faster and simpler to issue, and it is more readily amended or replaced. Without regulatory authority to enforce compliance, however, it may lack substance. Adaptation of existing laws, as was done in the USA, avoids the necessity of drafting new laws but can lead to redundancy and delays, for example, where a product such as Bt potatoes cannot be marketed until cleared by three separate regulatory agencies.

As a further consideration, there may be existing regulations governing importation of plants, animals, and other living organisms, sanitary and phytosanitary regulations, and quarantine rules and procedures. Operations of the biosafety system need to be coordinated with these other regulatory systems so that potential conflicts or confusion can be minimized.

The People

Applicants seeking to conduct field tests of GMOs and members of biosafety review committees who must make a decision to approve, or not, a proposed release share primary responsibility for ensuring the safe use of genetically engineered crops released into the environment through the processes of risk assessment and risk management. Additionally, in some biosafety systems, inspections of field test sites are made before, during and after the trial is conducted in order to verify compliance with biosafety standards and any further requirements imposed as a condition of approval.

All of these people–applicants, reviewers, and inspectors–need to be familiar with the environmental risk/benefit issues associated with biotechnology products and have a working knowledge of the biosafety review process and the role of risk management procedures. The quality of biosafety review and decision making, and the safe and appropriate handling of GMOs in the environment, is a direct outgrowth of their training and experience. Thus the challenge is to build a core of knowledgeable people who understand the issues and their potential consequences, and who can systematically evaluate the risks and benefits of a proposed GMO release.

The Process

A biosafety review systematically evaluates a GMO, the site where it will be released, and the conditions under which the release will be conducted. Inherent in the process is the idea that, just as with any technology or human endeavor, there is no such thing as zero risk.

Proposals to conduct field tests of GMOs are subject to review by a national biosafety committee (NBC) that assesses the safety of the proposed release by taking into account:

- the nature of the organism that receives the new trait;
- the donor organism from which the trait was derived;
- the vector or mechanism used to transfer the trait;
- the nature of the introduced trait, including potential toxicity of a gene product;

- characteristics of the site or environment into which the GMO will be introduced;
- elements of the release plan that provide containment or control.

Such tests, conducted under conditions of confinement in small plots or fields of limited size, do not present the same risks as do the large-scale, unconfined releases of commercial use.

Approval for commercialization typically requires another round of environmental biosafety review that considers, to the extent possible, potential effects associated with long-term and widespread use of the GMO. The review process for commercial use of a GMO begins after successful completion of several years of small-scale field tests (for newly developed varieties), or upon request from a company seeking to import GMO variety(ies) already in commercial use in the country of origin. If a potential risk of sufficient probability and having consequences of sufficient magnitude should be identified, the applicant determines suitable risk management measures that reduce the risk to an acceptable level.

Commercialization of GMOs intended for food or feed use are also subject to a food safety review, generally conducted by a committee under the Ministry of Health or its equivalent. When any and all biosafety considerations have been satisfied, the NBC makes a recommendation to the decision making authority. Final approval for commercial use may encompass nonscientific issues, such as accordance with quarantine regulations, international trade and treaty considerations, and socio-economic impact. It is during the process of approval for commercialization that seed registration requirements typically come into play.

The Public

Ultimately, the fate of agricultural biotechnology is in the hands of the public. It is the public's acceptance or refusal of LMOs and genetically modified foods that will determine whether or not the potential benefits of the technology are realized. Public acceptance depends on the extent to which consumers trust the biosafety regulatory agencies, and feel they have the necessary information to make informed decisions regarding the benefits and risks. Building such a public understanding should be a fundamental goal of any policy seeking to apply the new techniques of biotechnology to address constraints in agricultural productivity and sustainability.

Implementing biosafety policy does not end once guidelines are written, people are trained, and reviews are conducted. It is a dynamic process that evolves through mechanisms for incorporating new information. As reviewers' experience and familiarity with introductions of GMOs increase, flexible regulations allow a gradual relaxation of oversight for those releases that pose little or no risk to human health or the environment. Limited biosafety review resources can then be focused on cases that do present some element of risk.

Feedback mechanisms are the built-in supply lines through which guidelines, people, and the review process keep up with the rapid pace of scientific advance and an accelerating rate of small- and large-scale releases of GMOs worldwide. Access to technical information and data gathered from prior releases is essential to support and strengthen subsequent biosafety decisions. At the same time, the accumulated experience of biosafety reviewers constitutes another source of feedback that can be used to improve oversight procedures. In a well-designed system, scientific and procedural feedback operate simultaneously to improve the quality of biosafety decisions.

Scientific feedback comes from external as well as internal sources. In evaluating a proposed release, biosafety review committee members may benefit from the experience of other countries by considering the acceptability of data from similar releases conducted elsewhere. Databases of field test information and biosafety reviews from the USA, Europe, and Australia can be useful to developing nations, so long as potentially significant differences in the environment, affected ecosystems, and agronomic practices are recognized.

Review committees anticipating a need for data may specify reporting and record-keeping requirements as a condition of approval. Where appropriate, monitoring procedures specified as a condition of approval can be tailored to generate useful information. This type of feedback allows institutional and national biosafety review committees to gather information and identify emerging issues of local or regional importance.

The criteria for an effective biosafety system may be defined by asking a number of practical questions. Are applications reviewed on a timely basis? Is the burden of paperwork excessive or redundant? Are there new factors to consider, such as an administrative reorganization or a change in the budget? Is the review process transparent, flexible, and based on science? Do decisions have substance and weight? Is compliance the norm? Is the public satisfied that their health and safety concerns are being properly addressed? A periodic evaluation of regulations and implementation procedures gives applicants, reviewers, regulators, and the public an opportunity to assess how well the system is working and identify needed improvements.

BIOSAFETY SYSTEMS IN EGYPT AND ARGENTINA

ISNAR and Virginia Polytechnic Institute and State University (Virginia Tech) in 1999 set up a collaborative research project to assess the efficacy of biosafety systems in selected countries, by reviewing biosafety policies and procedures associated with the introduction and commercial use of GMOs. The project's objectives are:

- To assess the efficacy of biosafety policies and procedures associated with the introduction of biotechnology products in Argentina;
- To develop recommendations for enhancing the operation of Argentina's biosafety system and minimizing potential constraints to technology transfer;
- To identify areas where ISNAR and other international providers may offer further assistance.

Major points of interest in the study are: (1) the organization, membership, and operations of the government agencies involved in regulating GMOs; (2) the nature and availability of information on biosafety procedures and requirements; (3) the path of regulatory review and approval leading to commercial release; and (4) the personal experiences of applicants and reviewers in dealing with the biosafety system.

Egypt and Argentina were selected as the first two case-study countries, based on their extensive experience in agricultural biotechnology research, and in biosafety review of applications for GMO field trials (Egypt and Argentina) and commercial releases (to date, only in Argentina). In Egypt, in the period 1995-1999, 24 applications for contained or open field trials were reviewed by the NBC and 23 permits were issued. Three genetically-modified crops are moving toward commercial release. Argentina's biosafety system, established in 1991, was one of the earliest to be set up due to a combination of factors. First, agriculture is the country's strongest economic sector and the new technology was seen as a means of increasing production and therefore exports. Second, US and transnational seed companies were looking for locations where their local branches or affiliated growers could conduct "off season" trials, thereby accelerating the development of new varieties. Third, GMO research was well underway at several public research institutions, so qualified people with expertise in molecular biology and related disciplines were available for the task of developing a workable biosafety system. In the period 1991-1999, 367 applications for field trials were approved in Argentina. As shown in Table 1, genetically-modified maize, soybean and cotton are grown on a commercial scale.

The following sections describe the ways in which biosafety systems have been set up in the two countries.

Biosafety Regulations in Egypt

The Ministry of Agriculture formally instituted Egypt's national biosafety system and Land Reclamation (MALR) in two decrees issued in early 1995.

TABLE 1. Area of GM crops in Argentina

Crop/Year	Total Area (million ha)	GMO Area[1] (million ha)	% GMO
Soybeans			
96/97	6.67	0.05	0.8
97/98	7.18	1.4	19.5
98/99	8.4	6.1	73.0
99/00	8.5	6.8	80.0
Maize			
98/99[2]	3.27	0.03	0.9
99/00[3]	3.65	0.2	5.5
Cotton			
98/99	0.75	.005	0.7
99/00	0.34	.008	2.4

[1] Estimated
[2] Includes 7,000 h glufosinate tolerant maize and 23,000 h Bt maize
[3] Includes 5,000 h glufosinate tolerant maize and 195,000 h Bt maize

Source: Figures from DMA/ Granos–SAGPyA, cited in Burachick and Traynor (2001)

Ministerial Decree No. 85 (January 25, 1995) establishes a National Biosafety Committee (NBC); Ministerial Decree No. 136 (February 7, 1995) adopted biosafety regulations and guidelines for Egypt. The system involves several ministries, organizations and/or government agencies involved with the importation, exportation, and local production of natural products.

Applications to field test genetically-modified plant material are submitted to the Chair of the National Biosafety Committee. Genetically-modified material to be imported requires an import permit that must be obtained in advance from the Supreme Committee on Food Safety, Ministry of Health. Requests should be made a minimum of eight weeks prior to the proposed initiation of the importation or field test.

Procedures for commercializing GMO crops were established in 1998 by Ministerial Decree No. 1648. For varieties produced within Egypt, the process is as follows:

- The applicant completes a permit application form providing details of the genetic material introduced, the process used for inserting it, and other relevant information. The applicant also provides data from food and feed safety studies and evidence supporting a determination of low or negligible environmental risk. Where applicable, the applicant provides documents indicating approval of similar GMOs for release in their country of origin.

- The application form is submitted to the NBC, which after examination and approval forwards it to the Seed Registration Committee (SRC) for their preliminary approval to proceed with standard field trials conducted at several locations. The SRC assigns a team of qualified inspectors drawn from relevant ARC units and/or private certified laboratories to supervise cultivation, ensure adherence to any biosafety requirements, confirm the new phenotype, and evaluate agronomic performance.
- The NBC has the right to confirm the nature of the genetic modification by taking samples from the field for molecular analysis.
- After successful completion of the field trials and submission of a report to the NBC, the NBC authorizes the applicant to submit an application to the SRC for final approval to commercially release the new variety. Pending this, three years or seasons of agronomic performance trials are conducted under the supervision of the SRC.

The process for securing commercial release approval for crops genetically engineered outside of Egypt has an added step. The applicant must first obtain a permit for importing the initial seed material from the Supreme Committee for Food Safety, Ministry of Health. The permit is then presented to the NBC and the Seed Registration Committee, after which the seed is imported into the country. From this point forward, the remaining steps in the approval process are exactly the same as for GMOs developed within Egypt.

Biosafety Regulations in Argentina

In Argentina, the Agricultural Directorate within the Secretary of Agriculture, Livestock, Fisheries and Food (SAGPyA) regulates the use of GMOs in field tests, unconfined releases, and commercial applications. National biosafety guidelines in the form of non-legislative resolutions are part of the broader regulatory system governing the agricultural sector, in particular laws related to the protection of plant and animal health and seed registration. The Argentinean biosafety system encompasses four sets of guidelines that apply to the development and use of GMOs and their products: greenhouse research with transgenic plants, environmental release of plants and microbes for field tests and large-scale plantings, food safety, and the handling and confined release of transgenic animals.

Argentinean biosafety guidelines for release of genetically-modified plants and microorganisms are detailed in SAGPyA Resolutions No. 656/92, No.

837/93, and No. 289/97. They are based on the regulations and experience of the US and UK, adapted to local needs. CONABIA, the national biosafety committee, is the agency in charge of implementing the environmental release guidelines. The process leading to the commercial release entails three kinds of regulatory review:

- An environmental risk assessment conducted by CONABIA.
- A food safety evaluation conducted by the National Service for Agrifood Safety and Quality (SENASA)–the agency within SAGPyA with a mandate to regulate food safety and quality, animal health products (e.g., vaccines), and pesticides. A supporting Technical Advisory Commission (TAC) on the use of GMOs was created recently, to provide SENASA with an external, multidisciplinary advisory body that will give a broader base to its regulatory decisions.
- A further requirement for commercialization of GMOs comes under the National Directorate of Agrifood Markets (DNMA). The agency's review consists of an assessment of the possible impact of commercialization of the GMO may have on Argentina's international trade. DNMA has two divisions, the Directorate of Markets and the Directorate of International Affairs. The former is in charge of the market-oriented review, which comes after the environmental biosafety approval by CONABIA and food biosafety approval by SENASA.

CONABIA, SENASA, and DNMA reviews form the basis for a resolution prepared by CONABIA which, when signed by the Secretary of Agriculture, grants approval for commercial use of the GMO. Before actual sale, the applicant must apply to the National Institute of Seeds (INASE) for a New Variety Registration as required by regulations controlling proprietary and commercial practices. Where the GMO has pesticidal properties, such as plants expressing genes encoding Bt endotoxin proteins, commercialization requires specific authorization from SENASA for its use. The entire approval sequence leading to commercialization is diagrammed in Figures 1 and 2.

IMPLICATIONS FOR SEED REGULATORY AGENCIES

The above description of the biosafety systems in these two countries points to the additional steps needed for variety registration of genetically-modified seeds, and additional responsibilities required from national seed regulation agencies.

In Egypt, the seed registration system plays an equalizing role in the process leading to commercialization of agricultural GMOs. Seed registration in Egypt follows a well-established procedure controlled by several organizations under

FIGURE 1. Field Test Approval Procedure in Argentina

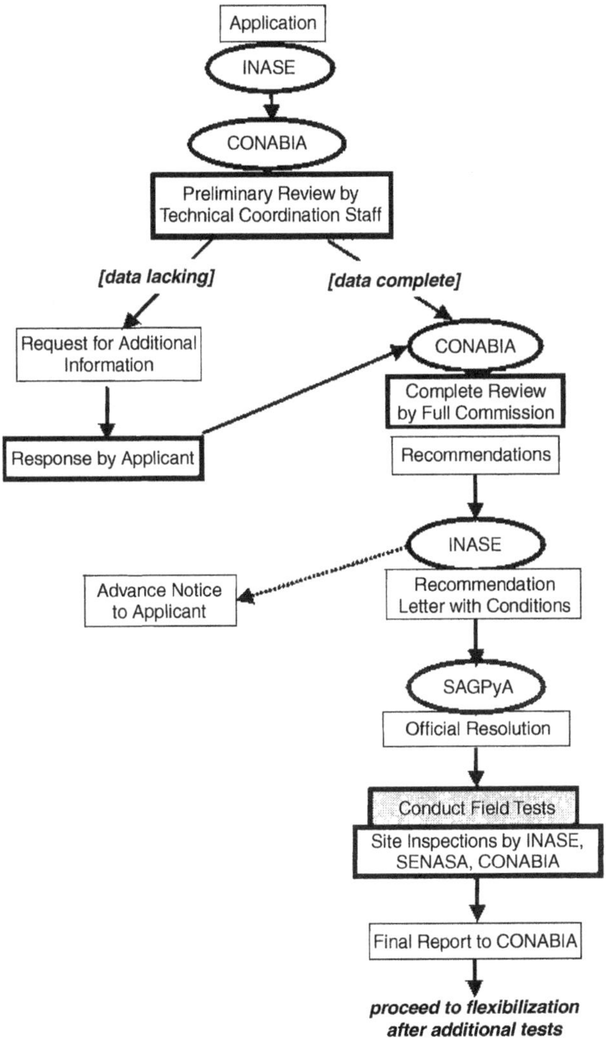

```
                    ┌─────────────┐
                    │ Application │
                    └─────────────┘
                           │
                       ( INASE )
                           │
                      ( CONABIA )
                           │
              ┌──────────────────────────┐
              │   Preliminary Review by   │
              │ Technical Coordination Staff │
              └──────────────────────────┘
                 │                    │
          [data lacking]        [data complete]
                 │                    │
     ┌─────────────────────┐          │
     │ Request for Additional │     ( CONABIA )
     │     Information     │          │
     └─────────────────────┘   ┌──────────────────┐
                 │             │  Complete Review  │
                 │             │ by Full Commission │
     ┌─────────────────────┐   └──────────────────┘
     │ Response by Applicant │  ┌──────────────────┐
     └─────────────────────┘   │  Recommendations │
                               └──────────────────┘
                                        │
                                    ( INASE )
                                        │
     ┌────────────────┐         ┌──────────────────┐
     │ Advance Notice  │┄┄┄┄┄┄┄│   Recommendation  │
     │  to Applicant   │        │ Letter with Conditions │
     └────────────────┘         └──────────────────┘
                                        │
                                   ( SAGPyA )
                                        │
                               ┌──────────────────┐
                               │ Official Resolution │
                               └──────────────────┘
                                        │
                               ┌──────────────────┐
                               │ Conduct Field Tests │
                               └──────────────────┘
                               ┌──────────────────┐
                               │ Site Inspections by INASE, │
                               │  SENASA, CONABIA  │
                               └──────────────────┘
                                        │
                               ┌──────────────────┐
                               │ Final Report to CONABIA │
                               └──────────────────┘
                                        │
                            proceed to flexibilization
                              after additional tests
```

the umbrella of the Ministry of Agriculture and Land Reclamation (MALR). The Central Administration for Seeds (CAS), an under-secretariat of MALR, has overall authority for policy and procedures regarding registration and release of all seed in Egypt. The standard testing procedure for registering conventional crops consists of trials conducted over three years (or three growing

FIGURE 2. Commercial Release Approval Procedure in Argentina

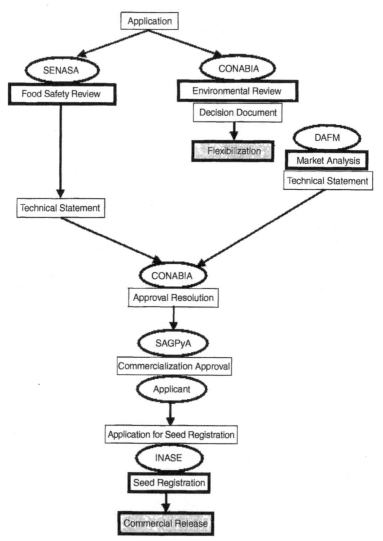

seasons) to document the uniformity and agronomic performance of new varieties. The same requirement applies to both locally developed and imported crops.

Once confined field tests of GMO varieities have been conducted, applicants submit a package to the National Biosafety Committee to support a finding of low or negligible environmental risk. Upon review and approval, the application is forwarded to the Seed Registration Committee (SRC), which grants approval to begin the same multiyear testing which is standard for all new varieties. Only upon successful completion of these trials will a GMO variety be approved for commercial sale. Thus seed registration is the final step in commercial release of genetically engineered crops.

In Egypt, the insect-resistant Bt maize lines approaching commercial release have raised a new issue regarding the SRC requirements. The Bt maize varieties were developed in parental lines that were already registered in Egypt. In the companies' view, addition of a Bt gene changes the phenotype to insect resistant, but does not change the line's other characteristics. They contend that GMOs derived from registered lines should not be subject again to the full testing scheme. The SRC has taken the position that insertion of a new gene makes the lines "new" and therefore subject to the full three-years testing.

Argentina's biosafety system follows a similar route to commercialization of new GM varieties. The National Institute of Seeds (INASE) is the agency within SAGPyA in charge of registration and control of commercially marketed seeds. Depending on the species, registration of new varieties requires two or three years of comparative field trials carried out at different locations. As in Egypt, transgenic varieties are seen as new hybrids, and have to undergo performance testing in comparison against the conventionally derived parent variety in the same way as new non-GMO varieties. Field trials of GMOs, made under conditions indicated by CONABIA after environmental safety assessment, are accepted as trials for registration purposes. A Technical Advisory Committee of the National Seed Commission (Comision Nacional de Semillas, CONASE) reviews the results and decides if the material qualifies as a new variety. Variety registration may be undertaken after a commercialization permit has been granted.

In the case of Argentina, the agency responsible for seed registration also functions as part of the biosafety system. INASE is the agency designated to receive and log applications for GMO field trials. INASE also communicates with applicants about matters regarding: (1) any deficiency detected by the Biosafety Commission in reviewing a field trial authorization request, and (2) any specific conditions which CONABIA has decided the petitioner should meet for the test to be carried out, based on biosafety considerations. INASE personnel perform field test site inspections, checking for compliance with the biosafety requirements set by CONABIA. Applications containing confidential business information are kept secure at INASE's offices.

RECOMMENDATIONS

As described in this article, the road to commercialization of genetically-modified varieties involves a wide range of government agencies dealing with environmental safety, food safety, seed registration, and, in the case of Argentina, market analysis.

The main challenge being faced in both countries is to build the necessary human capacity to conduct science-based risk assessments and biosafety studies. In the two countries studied, limited technical capacity in biotechnology creates the potential for conflicts of interest, when biosafety committee members also appear in the role of applicant seeking committee approvals. This situation has the potential to erode public confidence in the biosafety system unless the applicant officially removes himself from the review committee. Also, due to the range of agencies involved in approval procedures, there are some overlaps among the members of different committees, which may undermine the validity of independent reviews. The remedy for having individuals serve multiple functions lies in long-term capacity building to alleviate the shortage of technically qualified people.

In addition to biosafety review and approval, GMO crops intended for commercial use also are subject to existing regulations for seed registration, and may involve sanitary and phytosanitary regulations governing import, as well. There is potential for the respective authorities to overlap, which could lead to discord between the agencies and confusion on the part of the proponent. In both Egypt and Argentina, regulatory officials have worked to coordinate their respective requirements, with varying degrees of success. Such difficulties could be avoided by better coordination among agencies at the time biosafety implementation procedures are being worked out.

Particularly regarding seed regulatory agencies, procedures for streamlined or accelerated seed registration testing must be considered. There is no obvious scientific requirement to consider every GM plant variety as new, and therefore subject to several years of performance testing. Furthermore, the extent of biosafety data that can be derived from performance testing should not be overstated. Unless a trial includes a properly designed and controlled biosafety study, only observations can be made.

As in most countries, a major part of biotechnology research and development occurs in the private sector. Companies operate on a timeline for getting new products through each stage on the way to the market place, and thus for planning purposes need to know how long regulatory decisions will take. Government agencies involved directly or indirectly with the biosafety system would improve the regulatory environment by adhering to a set timetable for review and decision making. This way, they would avoid unnecessary delays during the application, review and approval process that may raise barriers to product development.

REFERENCES

Burachick, M.S. and P.L. Traynor. 2000. *Commercializing Agricultural Biotechnology Products in Argentina: Analysis of Biosafety Procedures.* Report prepared for ISNAR's Biotechnology Service. The Hague: International Service for National Agricultural Research.

Madkour, M., A.S. El Nawawy and P.L. Traynor. 2000. *Analysis of a National Biosafety System: Regulatory Policies and Procedures in Egypt.* ISNAR Country Report 62 (co-published with the Agricultural Genetic Engineering Research Institute). The Hague: International Service for National Agricultural Research.

Traynor, P. 1999. *Biosafety Management: Key to the Environmentally Responsible Use of Biotechnology.* Chapter 13 in Joel I. Cohen (Ed.) *Managing Agricultural Biotechnology: Addressing Research Program Needs and Policy Implications.* Wallingford: CABI Publishing.

An Agrobiodiversity Perspective on Seed Policies

Bert Visser

SUMMARY. The genetic erosion that followed the emergence of scientific plant breeding and the changes in agricultural production are the result of globalization and cannot be attributed to seed policies and legislation in particular. However, seed regulations can have a very important and often negative impact on local seed systems and the genetic diversity that is used and maintained in such systems. Also, seed legislation and intellectual property rights have a marked effect on formal and participatory plant breeding programs and on the number of varieties released to farmers. The recent developments of international regulations on intellectual property and the upcoming biotechnology revolution are likely to aggravate the current trend. Thus, policies on plant genetic resources and agrobiodiversity on the one hand and seed policies on the other hand influence each other and should be closely connected.

International agreements such as the FAO International Undertaking on Genetic Resources for Agriculture and the Convention on Biological Diversity may have a positive, regulating effect on the balance of power, but it is not yet clear whether these agreements will indeed have the desired effects on agrobiodiversity. Policy makers that develop or redesign seed policies and legislation should take international and local biodiversity issues and the objectives of the international agreements into account. *[Article copies available for a fee from The Haworth Document Delivery Service: 1-800-HAWORTH. E-mail address: <getinfo@haworthpressinc. com> Website: <http://www.HaworthPress.com> © 2002 by The Haworth Press, Inc. All rights reserved.]*

Bert Visser is affiliated with the Centre for Genetic Resources, The Netherlands (CGN), Plant Research International Wagenbingen, The Netherlands.

Address correspondence to: Bert Visser, Centre for Genetic Resoruces, The Netherlands (CGN), P.O. Box 16, 6700 AA Wageningen, The Netherlands.

[Haworth co-indexing entry note]: "An Agrobiodiversity Perspective on Seed Policies." Visser, Bert. Co-published simultaneously in *Journal of New Seeds* (Food Products Press, an imprint of The Haworth Press, Inc.) Vol. 4, No. 1/2, 2002, pp. 231-245; and: *Seed Policy, Legislation and Law: Widening a Narrow Focus* (ed: Niels P. Louwaars) Food Products Press, an imprint of The Haworth Press, Inc., 2002, pp. 231-245. Single or multiple copies of this article are available for a fee from The Haworth Document Delivery Service [1-800-HAWORTH 9:00 a.m. - 5:00 p.m. (EST). E-mail address: getinfo@haworthpressinc.com].

KEYWORDS. Genetic erosion, genetic resources, agrobiodiversity, benefit sharing, local seed systems

INTRODUCTION

Seeds form the subject of policies and their formalized version, legislation. Seeds are subjected to policies and legislation because they represent major values. Food production and food security are largely based on seeds, and in many countries seed production and seed supply involve major economic activities.

The capacity to produce food is, however, only one essential characteristic of seed. The fact that seeds are living material and, like all living materials, display diversity forms another essential property. This paper focuses on the genetic diversity encompassed in seeds, and in particular on the experienced and potential effects of seed policies on seed diversity. It also takes into account the diversity of human actors who deal with seeds and their diversity.

The design and revision of seed policies and intellectual property rights systems should accommodate for these effects on genetic diversity, and indirectly on food production, food security and cultural identity, and should recognise the roles of various stakeholders in the production of our food and its diversity.

Seed genetic diversity is regarded as part of agrobiodiversity. The function of agrobiodiversity in securing future food production and realizing more sustainable forms of agriculture is briefly discussed. How traditional seed supply systems developed into large global economic enterprises and how this affected the diversity of seeds are issues subsequently touched. Investments of the private seed industry require protection by intellectual property rights. The effects on acces to and utilisation of seed genetic diversity exerted by intellectual property rights form a central aspect of the relationship between seed policies and the management of agrobiodiversity. This includes:

- plant breeder's rights according to UPOV as an example of a *sui generis* property right system,
- patent rights, which invaded plant production from technology industry, and
- current efforts to develop an analogous system of farmer's rights play in this process brings us to the heart of the link between seed policies and management of agrobiodiversity.

It is questioned whether the current versions and uses of intellectual property rights follow from the need to promote inventions in biological and genetic properties of seeds, or whether they mainly follow from established economic interests.

SEEDS AS A CORNERSTONE OF AGROBIODIVERSITY

Various definitions of the terms biodiversity and agrobiodiversity have been formulated by policy makers and scientists. Here, widely accepted descriptions are reiterated. The Convention on Biological Diversity, agreed in 1992, defined biodiversity as "the variability among living organisms from all sources including, *inter alia*, terrestrial, marine and other aquatic ecosystems and the ecological complexes of which they are part; this includes diversity within species, between species and of ecosystems" (CBD, 1992, www.biodiv. org). Thus, this definition distinguishes three levels of integration with increasing complexity. Sublevels can be recognized within each level. Issues of diversity within species, i.e., genetic diversity, can be addressed at the species and population levels. The central criteria of distinctness of the UPOV treaty deals with genetic diversity at the species level between populations, whereas the criteria of uniformity relates to the within-population diversity. Traditional farming systems employ within-population diversity to obtain yield stability.

Genetic resources signify all materials containing genetic diversity of actual or potential value for food and agriculture, and the term most often directly refers to the diversity within species.

Agricultural biodiversity, also known as agrobiodiversity, forms a subset of total biodiversity. It refers to biodiversity related to agriculture and can be described as "the variety and variability amongst living organisms (of animals, plants, and microorganisms) that are important to food and agriculture in the broad sense and associated with cultivating crops and rearing animals and the ecological complexes of which they form a part." It includes the diversity found in farming systems as well as their surroundings to the extent that the latter influences agriculture.

However, these seemingly well-formulated definitions appear not all to be workable in practice. From a recent website discussion amongst the various stakeholders in the food production chain on options for an enlarged role for agrobiodiversity in Dutch agriculture it appeared that most participants had only a vague perception of agrobiodiversity, and consequently many participants disagreed in the value and role of agrobiodiversity in current agriculture (Pistorius et al., 2001). Such results reflect both the low degree of recognition of the value of agrobidiversity in agriculture as well as the complexity of the defined systems. Seed policies are often unconnected to (agro-)biodiversity policies. Seed policies heavily influence the development of production and seed supply systems. Whether agrobiodiversity can be equally sustained in strongly diverging production and seed systems or not, is explored to better understand how seed policies influence agrobiodiversity.

THER ROLE OF AGROBIODIVERSITY IN FOOD PRODUCTION

It is obvious that seeds play a major role in the maintenance of agrobiodiversity. At the within-species level, seeds contain the genetic diversity within a crop species, and are the vehicle of recombined genotypes and newly formed diversity. At the species level, availability of seeds determines the survival of neglected crops whose existence is threatened. And by influencing the cropping pattern, seeds influence the agro-ecosystems, in particular their environmental fitness at large and their sustainability. In other words, seeds are important at all three integration levels of agrobiodiversity.

But why is maintenance of agrobiodiversity important? In short, a high degree of agrobiodiversity in agro-ecosystems improves the buffering capacity and resilience of such systems when biological or climatic factors influence or alter these production systems: new pests and diseases may emerge, temperatures or rain fall patterns may gradually change. A high degree of agrobiodiversity means that humans keep access to resources that can help them to cope with such changing circumstances. When some crops or some varieties within crops fail because of lack of rainfall, others may survive and produce food. When new diseases occur, existing genetic diversity may be screened and exploited to detect resistances to such diseases. Consumer preferences might change due to urbanization, increased communication and exposure to other food habits. Foreign or forgotten food and crops might (re)gain importance and the demand for such crops might increase. Such developments require adaptation of production systems as well, and are dependent on the capacity of agro-ecosystems to respond to such shifting demands. A high degree of agrobiodiversity also decreases the dependence on high-external inputs, whether pesticides or chemical fertilizers. New pests can be better controlled by predators or antagonists present in the agro-ecosystems and green or animal manure may replace part of the need for chemical fertiliszers. In few words, agrobiodiversity has a major value for securing future food production and improved sustainability, but it also influences the capacity of the production systems to respond to short-term changes. Finally, it should be recognized that agrobiodiversity does not only reside in the crop but also in the knowledge in the use and properties of that crop. Indigenous knowledge is an integral part of agrobiodiversity.

Some additional remarks should qualify the notions mentioned above. The rather recent term "agrobiodiversity" only offers a new perspective on agricultural production factors. In fact, traditional agroecosystems have always relied on the principles contained in this concept, and modern western agriculture and seed industry may well be able to integrate the principles involved. Management and conservation of agrobiodiversity is not a goal in itself, but an instrument in achieving the goals behind it, securing food production and enhancing sustainability, in recognition of the fact that agrobiodiversity is only one factor to contribute to these goals. The use of the concept of agrobiodiversity resides in viewing farm production from a wider and integrated perspective, and in focussing on the relationships between various production factors and their environment. With regard to genetic resources, major efforts have been undertaken in the last few decades to conserve genetic diversity in genebanks, thus reducing our dependence on the survival of diversity in the field. However, it has become apparent that genebanks can only fulfil a limited role in conrsevation efforts, and on-farm conservation has been recognized as an important complementary strategy.

TRADITIONAL SEED PRODUCTION AND EXCHANGE

Until the end of the 19th century, everywhere in the world farmers produced seeds and consciously or unintentionally improved their crops, now known as farmer's varieties or landraces. Such improvement might simply stem from an on-going genetic adaptation of the variety to the agroecosystem. It might also involve a conscious selection, mass selection or pedigree selection, by farmers of germplasm with desired qualities. The term farmer's varieties better reflects the latter practice.

Farmer's varieties generally have some distinct features. First, they are adapted to the local circumstances under which they were developed. Such circumstances may be rather constant or they may vary greatly from year to year, and from field to field. At the one extreme the agroecosystems of the North-West European Plains and the Great Plains in the USA and to some extent irrigated lowland rice cultivation systems in Southeast Asia can be grouped, on the other extreme can be found the poor-soil, rain-dependent production systems at various altitudes in sub-Saharan Africa. Second, farmer's varieties often exhibit a considerable degree of genetic heterogeneity. It is precisely this heterogeneity which renders these varieties more flexible and capable to change in reaction to altering natural conditions. Third, because of this heterogeneity and the circumstances under which these varieties are grown, farmer's varieties are not stable. And finally, because they change over the years and are managed independently by different communities, they may or may not be regarded distinct from each other.

Not only the farmer's varieties themselves have some distinct features, this is also true for the way they are maintained and exchanged. In traditional farming systems food production mainly serves self-subsistence. The surplus can be exchanged through informal mechanisms between farmers, often under principles which are widely known as "common heritage of mankind." Seed is given away under the assumption that one day such gifts will be reciprocated. Both sides benefit because it increases their access to seed diversity and therefore the resilience of their production system. Often the surplus is also marketed through local distribution channels, involving middlemen. In all cases, the further use of the seed is free and no informal or formal property rights are recognized in such systems.

Farmer's varieties have almost disappeared from western and transition countries, where they mainly survive in the hands of hobbyists or in alternative production systems. But they are still dominant in many crops in tropical countries, in particular the non-staple crops. Similarly, local exchange mechanisms have almost vanished in western countries but are still dominant in tropical countries.

In conclusion, farmer's varieties form an inherent component of traditional small-scale farming systems. Farmer's varieties exhibit a relatively high degree of genetic variability, they are often not distinct, uniform and stable, they are exchanged freely, and property rights are foreign to such varieties. And, of course, seeds embody these varieties.

FROM TRADITIONAL TO MODERN SEED SUPPLY SYSTEMS

Concomitant with the industrialization of agriculture in western countries during the 20th century, variety development and seed production became an affair of specialists. Farmers who were skilled in selection and crossing of plants made their living of these skills and small breeding companies developed. Farmers who had access to good soils and good production circumstances in general specialized in seed production.

The industrialization of agriculture not only involved a gradual specialization in breeding from expert farmers to small specialized family-owned breeding companies. It should be stressed that in current traditional farming systems such specialization can be recognized as well. The industrialiszation exhibited many more features. Most family-owned breeding companies have gradually merged into a small number of large international breeding companies. Industrialiszation also involved:

- the creation of global markets in which a few crops dominate and form the main target of modern breeding efforts;
- the growing importance of food processing, in which a limited number of raw product materials are used to generate a large array of consumer

products (e.g., the dairy industry), and in which different crops can be used to generate the same consumer goods (e.g., vegetable oils); and

- the development of uniform standards applied to new crop varieties to allow for mechanical treatment and harvesting of crops, and to animal races to rear them under standardized conditions.

Along with these developments, the use of heterogeneous plant varieties decreased, and this was reflected in a decreased number of non-uniform varieties in the market. These are two aspects of genetic erosion.

Whereas these developments first occurred in industrialized countries from the turn of the 19th into the 20th century onwards, a similar event took place in the nineteen seventies and eighties with the Green Revolution. Again, thousands of local varieties of staple crops such as rice and wheat were replaced by high yielding varieties which had been developed by the international agricultural research centres (IARCs), notably IRRI and CIMMYT. Although these changes took place in different production systems, often still small-scale, again the effect of globalization resulted in improved food production depending on increased external inputs for a small number of varieties, soon covering large areas of tropical production systems.

In the last decades, public funding in agricultural research, including breeding, decreased sharply. In western countries public expenditure now focuses on pre-competitive fundamental crop research, whereas in many tropical countries, few staple crops which are essential for food security (mainly cereals) or crops which are major foreign exchange earners (e.g., sugar cane in Cuba, oil palm in Malaysia) receive most attention in public research. An estimated two-thirds of all investments in breeding are currently private (Pistorius and Van Wijk, 1999).

Following these developments, a few dominant transnational companies now cater for the needs for seeds and new varieties of farmers in western countries and large-scale farmers of export crops in tropical countries (maize, soybean), either directly or through joint ventures with local companies (Pistorius and Van Wijk, 1999). This does not mean that farmers return yearly to the seed companies to obtain their seed. Contrary to widely held views, in many countries including western countries the majority of farmers produce their own seed or obtain it from sales "over the fence" (western countries) or through local markets (tropical countries). Similarly, improvement of staple crops in tropical agriculture has become highly dependent on the products of the IARCs.

A salient feature of this modern international seed supply system is that it has produced uniformity, not because of UPOV requirements but because of economies of scale. Breeding companies could grow because markets extended when the production environment was adapted to the crop rather than the other way around. Such adaptation was achieved through the introduction of land management, and the use of external inputs to improve soil fertility and to protect crops from pests and diseases. So lesser varieties were sold in growing markets. But also, as explained above, the remaining varieties had to conform to uniform cultivation standards to allow mechanical management and harvesting. So these lesser varieties contained and exhibited less genetic diversity.

Again, some qualification of the statements above is necessary. An effect of the growing size and international character of breeding companies is their increased access to genetic diversity which may result in genetically improved varieties, and their increased financial resources to exploit this diversity. In particular, pest and diseases resistance genes from wild relatives of any source have been introduced in several crops (e.g., cereals, potato, vegetables). This tendency partly counteracts the loss of diversity due to uniformity requirements.

SEED POLICIES RESULTING FROM SEED TRADE

Seed trade has become commercialized and has become an international if not global affair. Seed policies and seed legislation serve various goals. They serve to guarantee quality standards (viability, identity) to farmers buying seeds which come under seed regulation, and they serve to guarantee other quality standards (identity, properties of the produce) to the food processing industry and to consumers. In a large number of countries various seed quality control measures and regulatory forms of variety registration as a prerequisite for marketing have been introduced. The scope of such measures may differ, and may involve major crops only, or a wide array of crops (e.g., Indonesia, Morocco, Uganda and the European Union), and seed certification or variety labeling only (USA). These regulations may outlaw farmer's varieties where still in use (e.g., for maize and sorghum in Zimbabwe, Cromwell and Van Oosterhout, 1999) or the reintroduction of older varieties which form a cultural heritage or are believed to better fit in organic practice (e.g., Britain, the Netherlands). Such regulations have become feasible because of an increased dependence of farmers on large-scale seed markets: through these seed markets national governments were able to interfere.

But plant breeding and seed trade under conditions of modern agriculture required yet another form of protection: intellectual property rights.

Investments by private breeding companies to develop new crop varieties can only be justified if returns on these investments can be realized. A seed trade and production system which allows any third party, whether competing breeding or seed company or a large number of farmers, to grow and market the seeds of new improved varieties is incompatible with such private breeding investments. Yet, free exchange and use had been the rule and seed legislation was needed to change the rules, since biological means of protection through the development of hybrid varieties were only partially successful and not amenable for all crops.

Plant breeder's rights were introduced in the international arena though the UPOV convention, agreed in its first version of 1961 (Ghijsen, this volume). From a diversity perspective both the breeder's exemption and farmer's privilege are important provisions, allowing the further use of protected varieties for breeding, selection and adaptation. Any measure to limit the scope of these provisions is potentially detrimental to on-farm crop development.

As a new development and consequence of the growing role of plant biotechnology in breeding, patent rights have now also entered the scene. Patent rights originate from industry and were developed to fit the needs for protection of industrial processes and products. Patent rights do not fit breeding and will harm the interests of many stakeholders in the food production chain. In contrast to plant breeder's rights under the UPOV convention they do not allow seed production of protected varieties and their use for further breeding by third parties, whether smaller local companies or farmers (Visser and Engels, 2000). Therefore, the introduction of patent rights in plant breeding will have a negative effect on genetic diversity.

EFFECTS ON SEED DIVERSITY

To what extent have the changes in seed supply systems and subsequent seed policies and seed legislation influenced seed diversity?

The industrialiszation of agriculture in western agriculture resulted in a small number of major players. These players together employ less breeders in less countries with more limited local networks compared to the previous phase in which small-scale family enterprise was the dominant form, let alone the traditional organization in which innumerable farmers carry out selection and more advanced forms of breeding. By necessity, the number of varieties was gradually reduced and the genetic diversity in these varieties decreased because breeders drew from less sources. Far from a conspiracy, these developments were the results of the economies of scale.

But other stakeholders also influenced diversity. Since food trade has become internationalized and consumer habits have changed as a result of exposure to new food, the diversity of crops consumed after industrial or home processing has also decreased. Here, the seed market and industrial seed producers are certainly not the only factor in changing the seed supply. A change in food demands by consumers has also changed the seed supply.

In addition, the products of the Green Revolution have often been strongly promoted by national governments through linking loans and other benefits to the cultivation of these varieties. Seed policies by these governments were understandably driven by the prospects of high yields for staple crops and the need to feed a fast growing population.

In the FAO State of the World (1996) the Republic of Korea reports that 74% of varieties of 14 crops grown on particular farms in 1985 had been replaced in 1993. China reported that nearly 10,000 wheat varieties were used in 1949, but only 1,000 in the 1970s. Fifteen million hectares of hybrid rice in China share a common cytoplasmic male sterility source. Till 1970, about 5,000 varieties of rice were grown in India, but currently about 500 varieties are grown of which 10-20 may be covering a large part of the country. Only 20% of local maize varieties in Mexico reported in 1930 are still known today, although it should be noted that loss of varieties does not necessarily mean loss of diversity, since this can be conserved in new varieties. Although an estimated 7,000 other plant species have been used as food by humans at some time, approximately 60% of global food supplies are now provided by rice, wheat and corn, and 90% by a total of thirty crops only.

In addition to factors mentioned above, legislation on seed quality control and variety registration as an instrument to support increased crop production has frustrated efforts to generate or maintain varieties in local farming systems, to develop local seed enterprises, or simply to cultivate the best adapted varieties. The best-adapted varieties may not conform to DUS standards. Alternatively, local enterprises that would supply niche-markets with specifically adapted materials may not have the resources to allow seed inspection, or government agencies may not have the resources to select and list the large numbers of varieties of specifically adapted varieties. In particular, activities employing participatory variety selection (PVS) and participatory plant breeding (PPB) have suffered from interference by legal provisions. Examples are the prohibition of open-pollinated maize varieties in Zimbabwe (Louwaars & van Marrewijk, 1996), the selection of optimal rice and chickpea varieties in India (Witcombe et al., 2000), and the breeding of new bean varieties in Uganda (Louwaars, pers. comm.).

Even if seed legislation is not fully responsible for these developments, it definitely has strengthened the effects. In the current 1991 version of the UPOV convention, the farmer's privilege has been limited and sales across the fence are no longer allowed. Maybe more important is that as a result of the WTO Trade-Related Intellectual Property Rights (TRIPs) agreement the number of countries which have enacted UPOV-compatible and similar laws has drastically increased. Since informal marketing of protected seeds by local producers is thus increasingly controlled, the distribution of protected seeds and the subsequent incorporation of their preferred traits into local varieties may slow down. Alternatively, protection of these seeds and the higher prices requested by official traders may not only limit their use, but also divert attention to local varieties with their own variability which are freely available.

In conclusion, seed policies other than intellectual property rights seem to have had the most profound impact on genetic diversity in the field. This picture might change under the influence of biotechnological innovations in plant breeding.

ALTERNATIVES RECOGNIZING THE NEED FOR DIVERSITY

Several alternative strategies and corrective measurements have been proposed in the course of the last decade. Some of these corrective measures are modest. For example, it has been proposed that for countries which wish to join the UPOV convention as new members, options should be created to enter on the conditions of the earler versions of the convention, which sets more flexible conditions for the breeder's privilege and in particular the farmer's exemption. It has also been suggested to restrict the coverage of national legislation following a UPOV membership to a few staple crops and not to include an array of smaller crops which often exhibit higher levels of variability. In fact, this is a common feature of a number of seed laws; for example, Bangladesh has enacted restrictive seed regulations covering only five major crops. In general, seed regulations tends to discriminate between crops in contrast to IPR legislation. Another suggested corrective measure is to exclude traditional varieties from the need to register these as varieties, and to allow modest trade in such varieties. A proposal for EU legislation along these lines has been forwarded by Germany and is still being debated.

Still another corrective measurement is to revisit and revise the often strict criteria for variety registation and certification allowing a recognition of traits of particular importance for organic production. In organic production a yield penalty may be acceptable if counterbalanced by better resistance, better rooting and soil coverage, and improved taste. A "green" variety list has been suggested to accommodate the specific requirements of organic agriculture (e.g., in the Netherlands). Such a development would widen the genetic diversity effectively used in the field.

Finally, it has been suggested to remove the obligatory character from current seed legislation and to make variety registation voluntary or to make the national variety lists a recommended list only. Such a change has been suggested for sorghum and maize in Zimbabwe, as well as for the variety registration system on arable crops in the EU.

The common feature of each of these initiatives is that they are corrective: they accept the principles of current seed policies and only wish to decrease the negative effects. They are proposed and supported by different stakeholders:

government authorities wish to limit their involvement, the seed industry wishes to replace the obligatory status of seed variety registration by voluntary because of the high costs and lengthy time schedules involved for crops for which biological property protection is available (in particular hybrids), whereas small-scale farmers, organic farmers and hobbyists seek to widen their options to grow and market other varieties than the latest recommended high-yielding ones from the seed industry. In other words, whereas not all stakeholders seek to promote diversity, the net effect is probably that a wider diversity can develop in the field.

A more fundamental question to be asked is whether the current provisions of plant breeder's rights under the UPOV convention really stem from genetic necessity or whether they are the result of economic considerations from a time before the introduction of biotechnology in general and of molecular genetic markers in particular (Visser, 1998). The elements at stake are those of distinctness, uniformity and stability. Distinctness of a new variety from existing cultivars should be apparent through a sufficiently different and phenotypically expressed trait. Accepting a molecular difference at the genotypic level might form an alternative and complementary approach, rendering additional germplasm available through market mechanisms for further breeding. (The characters that determine DUS have nothing to do with agronomic value.) The requirement for uniformity somehow follows from the requirement for distinctness. If legislation requires distinctness between varieties it should create sufficient "room" for new varieties by limiting the genetic coverage of registered varieties. Indeed, this is an argument for uniformity which stems from economic but not from biological principles. One might argue that uniformity also serves the farmer in economizing on cultivation measures and improving the price for his produce, as well as the other players in the production chain. However, this argument is mainly valid for current large-scale modern production, and not for the organic sector, the small-scale sector, and probably other resource-intensive production sectors. Molecular markers offer options to relax the current rigid requirements for uniformity. A certain level of heterogeneity might be acceptable as long as this can be monitored and per variety documented through the use of molecular makers. Under this scenario populations of different varieties might even contain genetically very similar individual genotypes: what counts would be the overal molecular pattern of the entire variety. Relaxation of the requirement of uniformity might benefit the organic and small-scale production sector for which genetic variation might contain the necessary buffering capacity in production and the genetic make-up required for farmer breeding. In this respect, the current trend amongst maize breeders to introduce a strict interpretation of the provision on dependency between varieties (see Ghijsen, this volume) may be challenged. It is obvious that stability should also be reinterpreted: genetically mixed populations might conform to new stability standards which allow a change in frequencies of individual genotypes constituting the variety. In this respect, it should be stressed that even many of the varieties which pass the much heralded DUS requirements of plant breeders' rights appear not so genetically uniform as they appear phenotypically, when studied using molecular markers. This means that relaxing current DUS standards would not be an absolute but only a quantitative change in interpretation of the variety protection legislation.

THE MENACE OF PATENT RIGHTS AND TRANSGENIC CROPS

The number of marketed transgenic varieties is still limited, but the investments in their development by seed industry are substantial and it can be expected that a growing number of transgenic varieties will appear in the market in near future, regardless the exact size of their market shares. Under patent rights not only the marketing of protected varieties is forbidden. Genetically modified crops can be protected by patents on the introduced transgenes and the patent legislation does not allow for breeding and free regrowth of the seed. A first issue to be solved should be scope of the patent protection of a transgene. Is it the DNA sequence and the value of the trait in its particular (genetic) environment only, or can a patent on a gene and the trait also cover the use of that gene in traditional cross breeding? Undoubtedly, the latter interpretation would severely hamper crop improvement and limit genetic variability in future crop varieties through the privatization and exclusion of specific DNA sequences.

Patent protection on specific transgenes will mean that genetically-modified crops may no longer be used for small-scale breeding efforts by small breeding companies or farmers, and that farmers who grow crops in which these transgenes occur, consciously or unwillingly introduced, trespass legal provisions. The net effect of the introduction of patent rights in breeding will be that breeding with any genetic material is no longer a universal right and this may negatively influence further crop improvement, small-scale and large-scale alike. And by consequence, the development and long-term sustainability of local genetic diversity may be hampered. This is not a theoretical scenario, as several authors (e.g., Louette, 2000) have documented cases in which small-scale farmers had used commercial varieties to introgres desired traits from commercial varieties into their local germplasm.

From this perspective plant breeders rights systems should be strengthened to withstand their gradual replacement in breeding by patent rights, and such strengthening might also be achieved by removing some of the disadvantages of the current interpretations of plant breeder's rights under the latest UPOV Convention.

FARMERS' RIGHTS AND SEED POLICIES

The FAO International Undertaking on Plant Genetic Resources for Food and Agriculture has first described farmers' rights as the rights of farmers and farming communities to manage and develop, and benefit from their *plant genetic resources*. Several groups have subsequently proposed farmers' rights as a legal counterpart of plant breeder's rights. However, implementation of this principle encounters many obstacles, including the absence of clearly defined groups of legal right holders, the different variety concept used by small-scale farmers, and last but not least the culturally foreign concept of such a property regime to most farming communities. It can also be questioned if such a legal interpretation of farmers' rights would not harm the optimal exchange and utilization of diversity stemming from farmers' varieties. However, an economic interpretation of the concept, according to which communities which manage on-farm crop and genetic diversity are supported to continue such practices, might thus be beneficial to the survival of genetic diversity on-farm, and not only in the hands of breeders and genebanks.

In addition, the CBD recognizes the role of indigenous and local communities and agrees to "respect, preserve and maintain knowledge, innovations and practices of indigenous and local communities embodying traditional lifestyles relevant to the conservation and sustainable use of biological diversity and promote their wider application." This provision of the CBD reconfirms the concept of farmers' rights formulated in the FAO International Undertaking in a wider context and supports approaches to retribute and support farming communities maintaining genetic diversity and the associated knowledge in the form of concrete measures.

At the very least, seed policies should not be detrimental to efforts to maintain and develop crop genetic diversity. But even more important is that seed policies should encompass measures safeguarding the maintenance of genetic diversity on-farm and of agrobiodiversity, thus contributing to future food security and a more sustainable agriculture.

REFERENCES

Anon. 1983. *FAO International Undertaking on Plant Genetic Resources for Food and Agriculture.* FAO, Rome, Italy.

Anon. 1992. *Convention on Biological Diversity.* UNEP. (http://www.biodiv.org/)

Anon. 1996. *Report on the State of the World's Plant Genetic Resources for Food and Agriculture.* FAO, Rome, Italy.

Cromwell, E. and S. Van Oosterhout, 1999. On-farm conservation of crop diversity: policy and institutional lessons from Zimbabwe. *In:* S.B. Brush (Ed.) *Genes in the*

Field. On-Farm Conservation of Crop Diversity, pp. 217-238. Lewis Publishers, Boca Raton, USA.

Louette D., 1999. Traditional management of seed and genetic diversity: what is a landrace? *In*: S.B. Brush (Ed.) *Genes in the Field. On-Farm Conservation of Crop Diversity*, pp. 109-142. Lewis Publishers, Boca Raton, USA.

Louette, D. & M. Surale, 2000. Farmers' seed selection practices and traditional maize varieities in Cuzalapa, Mexico. Euphytico. 113: 25-41.

Louwaars, N.P. & G.A.M. van Marrewijk, 1996. *Seed Systems in Developing Countries*. Wageningen, CTA, 166 p.

Pistorius, R. and J. Van Wijk, 1999. *The Exploitation of Plant Genetic Information: Political Strategies in Crop Development*. CABI Publishing, London and New York, 231 pp.

Pistorius, R., N.G. Röling and B. Visser, 2000. Making agrobiodiversity work: results of an on-line stakeholder dialogue (OSD) in the Netherlands. Neth J Agric Sci 48: 319-340.

Visser, B., 1998. Effects of biotechnology on biodiversity. Biotechnology and Development Monitor 35: 2-6.

Visser, B. and J. Engels, 2000. The common goal of conservation of genetic resources. *In:* C. Almekinders and W. de Boef (Eds.) *Encouraging Diversity. The Conservation and Development of Plant Genetic Resources*, pp. 145-153, Intermediate Technology Publications Ltd., London.

Witcombe, J., K.D. Joshi, R.B. Rana and D.S. Virk, 2000. Participatory varietal selection and genetic diversity in high-potential rice areas in Nepal and India. *In:* C. Almekinders and W. deBoef (Eds.) *Encouraging Diversity. The Conservation and Development of Plant Genetic Resources*, pp. 203-207, Intermediate Technology Publications Ltd., London.

Index

T - #0503 - 101024 - C0 - 212/152/15 - PB - 9781560220930 - Gloss Lamination